W9-CJM-250

Qualitative Methods in Military Studies

This book examines the methodology of qualitative research in military studies.

Since the end of the Cold War, the number of studies on military and society has grown substantially in substance, size and impact. However, only a tiny part of this bibliography deals in-depth with the research methods used, especially in relation to qualitative methods. The data that forms the basis of the researchers' analyses are often presented as if they were immediately available, rather than as a product of interaction between the researcher and those who participated in the research.

Comprising essays by international scholars, the volume discusses the methodological questions raised by the use of qualitative research methodology in military settings. On the one hand, it focuses on the specificity of the military as a social context for research: the authors single out and discuss the particular field effects produced by institutional arrangements, norms and practices of the military. On the other, the authors proceed in an empirical manner: all methodological questions are addressed with regard to concrete situations of field research.

This book will be of much interest to students of military studies, research methods, sociology, anthropology, war and conflict studies and security studies in general.

Helena Carreiras is professor at ISCTE-University Institute of Lisbon and senior researcher at CIES-IUL. She was deputy director of the Portuguese National Defence Institute between 2010 and 2012, and holds a PhD in Social & Political Sciences from the European University Institute. She is the author of *Gender and the Military* (Routledge, 2006).

Celso Castro is professor and current director of CPDOC (The School of Social Sciences and History) at Getulio Vargas Foundation (Rio de Janeiro, Brazil). He has a PhD in Social Anthropology and is the author of several books on the military.

Cass Military Studies

Intelligence Activities in Ancient Rome
Trust in the Gods, But Verify
Rose Mary Sheldon

Clausewitz and African War
Politics and Strategy in Liberia and Somalia
Isabelle Duyvesteyn

Strategy and Politics in the Middle East, 1954–60
Defending the Northern Tier
Michael Cohen

The Cuban Intervention in Angola, 1965–1991
From Che Guevara to Cuito Cuanavale
Edward George

Military Leadership in the British Civil Wars, 1642–1651
'The Genius of this Age'
Stanley Carpenter

Israel's Reprisal Policy, 1953–1956
The Dynamics of Military Retaliation
Ze'ev Drory

Bosnia and Herzegovina in the Second World War
Enver Redzic

Leaders in War
West Point Remembers the 1991 Gulf War
Frederick Kagan and Christian Kubik (eds)

Khedive Ismail's Army
John Dunn

Yugoslav Military Industry 1918–1991
Amadeo Watkins

Corporal Hitler and the Great War 1914–1918
The List Regiment
John Williams

Rostóv in the Russian Civil War, 1917–1920
The Key to Victory
Brian Murphy

The Tet Effect, Intelligence and the Public Perception of War
Jake Blood

The US Military Profession into the 21st Century
War, Peace and Politics
Sam C. Sarkesian and Robert E. Connor, Jr. (eds)

Civil-Military Relations in Europe
Learning from Crisis and Institutional Change
Hans Born, Marina Caparini, Karl Haltiner and Jürgen Kuhlmann (eds)

Strategic Culture and Ways of War
Lawrence Sondhaus

Military Unionism in the Post Cold War Era
A Future Reality?
Richard Bartle and Lindy Heinecken (eds)

Warriors and Politicians
U.S. Civil–Military Relations under Stress
Charles A. Stevenson

Military Honour and the Conduct of War
From Ancient Greece to Iraq
Paul Robinson

Military Industry and Regional Defense Policy
India, Iraq and Israel
Timothy D. Hoyt

Managing Defence in a Democracy
Laura R. Cleary and Teri McConville (eds)

Gender and the Military
Women in the Armed Forces of Western Democracies
Helena Carreiras

Social Sciences and the Military
An Interdisciplinary Overview
Giuseppe Caforio (ed.)

Cultural Diversity in the Armed Forces
An International Comparison
Joseph Soeters and Jan van der Meulen (eds)

Railways and the Russo-Japanese War
Transporting War
Felix Patrikeeff and Harold Shukman

War and Media Operations
The US Military and the Press from Vietnam to Iraq
Thomas Rid

Ancient China on Postmodern War
Enduring Ideas from the Chinese Strategic Tradition
Thomas Kane

Special Forces, Terrorism and Strategy
Warfare By Other Means
Alasdair Finlan

Imperial Defence, 1856–1956
The Old World Order
Greg Kennedy

Civil-Military Cooperation in Post-Conflict Operations
Emerging Theory and Practice
Christopher Ankersen

Military Advising and Assistance
From Mercenaries to Privatization, 1815–2007
Donald Stoker

Private Military and Security Companies
Ethics, Policies and Civil-Military Relations
Andrew Alexandra, Deane-Peter Baker and Marina Caparini (eds.)

Military Cooperation in Multinational Peace Operations
Managing Cultural Diversity and Crisis Response
Joseph Soeters and Philippe Manigart (eds.)

The Military and Domestic Politics
A Concordance Theory of Civil-Military Relations
Rebecca L. Schiff

Conscription in the Napoleonic Era
A Revolution in Military Affairs?
Donald Stoker, Frederick C. Schneid and Harold D. Blanton (eds.)

Modernity, the Media and the Military
The Creation of National Mythologies on the Western Front 1914–1918
John F. Williams

American Soldiers in Iraq
McSoldiers or Innovative Professionals?
Morten Ender

Complex Peace Operations and Civil Military Relations
Winning the Peace
Robert Egnell

Strategy and the American War of Independence
A Global Approach
Donald Stoker, Kenneth J. Hagan and Michael T. McMaster (eds.)

Managing Military Organisations
Theory and Practice
Joseph Soeters, Paul C. van Fenema and Robert Beeres (eds.)

Modern War and the Utility of Force
Challenges, Methods and Strategy
Jan Angstrom and Isabelle Duyvesteyn (eds.)

Democratic Citizenship and War
Yoav Peled, Noah Lewin-Epstein and Guy Mundlak (eds.)

Military Integration after Civil Wars
Multiethnic Armies, Identity and Post-Conflict Reconstruction
Florence Gaub

Military Ethics and Virtues
An Interdisciplinary Approach for the 21st Century
Peter Olsthoorn

The Counter-Insurgency Myth
The British Experience of Irregular Warfare
Andrew Mumford

Europe, Strategy and Armed Forces
Towards Military Convergence
Sven Biscop and Jo Coelmont

Managing Diversity in the Military
The Value of Inclusion in a Culture of Uniformity
Daniel P. McDonald and Kizzy M. Parks (eds.)

The US Military
A Basic Introduction
Judith Hicks Stiehm

Democratic Civil-Military Relations
Soldiering in 21st-Century Europe
Sabine Mannitz (ed.)

Contemporary Military Innovation
Between Anticipation and Adaption
Dmitry (Dima) Adamsky and Kjell Inge Bjerga (eds.)

Militarism and International Relations
Political Economy, Security and Theory
Anna Stavrianakis and Jan Selby (eds.)

Qualitative Methods in Military Studies
Research Experiences and Challenges
Helena Carreiras and Celso Castro (eds.)

Qualitative Methods in Military Studies

Research experiences and challenges

Edited by Helena Carreiras and Celso Castro

LONDON AND NEW YORK

First published 2013
by Routledge
2 Park Square, Milton Park, Abingdon, Oxon OX14 4RN

Simultaneously published in the USA and Canada
by Routledge
711 Third Avenue, New York, NY 10017

Routledge is an imprint of the Taylor & Francis Group, an informa business

British Library Cataloguing in Publication Data
A catalogue record for this book is available from the British Library

Library of Congress Cataloging in Publication Data
Qualitative methods in military studies: research experiences and challenges/edited by Helena Carreiras and Celso Castro.
p. cm.
Includes bibliographical references and index.
1. Armed forces. 2. Sociology, Military–Methodology. 3. Sociology, Military–Case studies. I. Castro, Celso, 1963–
UA15.Q35 2013
355–dc23

2012008149

ISBN: 978-0-415-69811-5 (hbk)
ISBN: 978-0-203-09922-3 (ebk)

Typeset in Baskerville
by Wearset Ltd, Boldon, Tyne and Wear

Printed and bound in the United States of America by Publishers Graphics, LLC on sustainably sourced paper.

Contents

Contributors

Ana Alexandre holds a master's degree in Sociology from ISCTE–University Institute of Lisbon and is presently a research fellow at CIES–IUL, where she worked as a research assistant on a project on the role of the Portuguese Armed Forces after the Cold War.

Helena Carreiras holds a PhD in Social & Political Sciences from the European University Institute (Florence, Italy). She is a professor at ISCTE–University Institute of Lisbon and senior researcher at the Centre for Research and Studies in Sociology (CIES–IUL). Between 2010 and 2012 she was deputy director of the Portuguese National Defence Institute.

Celso Castro holds a PhD in Social Anthropology (National Museum/UFRJ, Brazil, 1995). He is professor and current director of CPDOC (The School of Social Sciences and History) at Getulio Vargas Foundation (Rio de Janeiro, Brazil). He is the author of several books on the military in Brazilian society and history.

Delphine Deschaux-Beaume holds a PhD in Political Science from the Institute for Political Studies of Grenoble (France). She teaches at several universities: IEP Grenoble, University of Savoie, University of Lyon 2 and IEDES-Paris 1-Sorbonne. She is currently research fellow at PACTE-CNRS and Ecole de la Paix de Grenoble.

Saïd Haddad holds a PhD in Political Science, and is Senior Lecturer in Sociology at Ecole Spéciale Militaire of Saint-Cyr, France. He is the head of the Department of Information & Communication and also a researcher at the Laboratory of Anthropology and Sociology at the University of Rennes 2 and at the Institute for Research and Study on the Arab and Islamic World (IREMAM/CNRS, Aix-en-Provence, France). His research interests include: minorities and ethnicity in Western armies; professionalisation and changes in military culture; culture as an operational stake for French armies.

Jelena Juvan holds a PhD in Defence Science (2008). She works as a graduate research fellow at the Defence Research Centre, Faculty of Social

Sciences, University of Ljubljana, Slovenia. Her areas of research are military families, war and religion, peace operations, the human factor in the military organisation, relations between military organisation and society, civil–military relations.

Charles Kirke had a full career in the British Army (36 years) and is currently Lecturer in Military Anthropology and Human Factors at the Centre for Human Systems, Cranfield University. His main research area is the organisational culture of the British Army.

Dirk Kruijt was Professor of Development Studies from 1993 to 2008 and is at present a Professor Emeritus at Utrecht University, the Netherlands. He was the former president of the Netherlands Association of Latin American and Caribbean Studies (NALACS) (1994–1998). His research interests are (economic, social and political) informalisation, armed actors, civil–military relations, military governments, organised crime and urban violence.

Phil C. Langer holds PhDs in Literary Studies (2002) and Psychology (2009) from Munich University. He is Assistant Professor of Sociology and Social Psychology at Frankfurt University. Between 2009 and 2011 he worked as a senior researcher at the Bundeswehr Institute of Social Sciences, Strausberg, and as a lecturer in sociology at Potsdam University. His areas of interest include military-related issues, health and gender issues, qualitative methodology in social research and Holocaust Education.

Piero C. Leirner holds a PhD in Social Anthropology from the University of São Paulo (2001) and is currently an associate professor at the Federal University of São Carlos. His work has emphasis on the anthropology of war, mainly in the following subjects: state, hierarchy, war and military.

Ian Liebenberg holds a PhD in Sociology, and an MA in Development Studies and Political Science. He is a senior researcher at the Centre for Military Studies (CEMIS), South African Military Academy, Saldanha, South Africa. He also lectures in the Department of Political Science, Faculty of Military Science at Stellenbosch University.

Alejandra Navarro holds a PhD in Social Science from the University of Buenos Aires. She is a lecturer in Qualitative Methodology at the University of Mar del Plata, and in Research Methods in the Social Sciences at the University of Buenos Aires.

Carsten Pietsch holds an MA in Political Sciences, Sociology and History (2006), and is developing a PhD project on decision-making bodies in the European Security and Defence Policy's architecture. He works as a researcher at the Bundeswehr Institute of Social Sciences in Strausberg and as a lecturer in Sociology at the University of Potsdam.

Janja Vuga graduated in political science from the Defence Studies Department of the Faculty of Social Sciences at the University of Ljubljana, Slovenia, where she is presently a doctoral candidate. Since 2007, she has been working as a researcher and teaching assistant at the Faculty of Social Sciences, Chair of Defence studies. Her areas of research are multinational peace operations, public opinion, civil–military relations and gender in the military.

Introduction

Celso Castro and Helena Carreiras

There is a large and rapidly growing body of military studies in the area of Social Sciences. A survey done by Kurt Lang, almost 50 years ago, of studies carried out in the U.S. and Europe, recorded a significant 528 texts. More recently, particularly after the end of the Cold War, the number of studies on military and society has grown substantially in substance, size and impact.

However, only a tiny part of this bibliography deals in-depth with the research methods used, especially in relation to qualitative methods. The data that form the basis of the researchers' analyses are often presented as if they were immediately available, rather than as a product of interaction between the researcher and those who, along with the military institution, participated in the research and made it possible.

This book aims to discuss issues related to this interaction, on the basis that this is essential in order to understand not only conditions in which research may have been done, but also the very nature of the data presented and of the final analysis. In other words, we assume that the conditions in which data is obtained are relevant to its very understanding. We believe that the researcher's integration into the research context – in this case, the military institution – affects the process of obtaining data and, as a result, its analysis.

How do researchers gain access to the military institution? What type of interaction do they establish with military personnel? How do they deal with the hierarchical world of the barracks? What happens when the researcher is a member of the military, or works for the institution as a civil servant? To what extent can gender differences affect interaction during research? How is the publication of the results received by the institution? These questions are approached from different points of view in the chapters that follow.

The aim of this book is, therefore, much less to present the content and specific results that each author obtained in their work (although these are also to some extent necessarily present in the texts) and more to reflect on the conditions under which qualitative research methods are used and how they are carried out in the context of the military institution.

A crucial decision made by the editors was to only select studies that discuss the use of qualitative methods referring to concrete empirical research experiences, since a discussion on research methods removed from its empirical contextualization cannot be fruitful.

We have privileged diversity, bringing together a set of contributions which deal with research that has used a range of qualitative methods and techniques – from ethnography and participant observation, to in-depth interviews or focus groups – as well as comparative and mixed-methods designs. We have also tried to consider researchers from different disciplinary backgrounds – Anthropology, Sociology, History, Social Psychology and Political Science – and various theoretical orientations, on the basis that this diversity enriches the discussion we wish to encourage. Lastly, the authors brought together here are from different countries, avoiding the risk of privileging overly ethnocentric perspectives on these issues. Researchers from Argentina, Belgium, Brazil, France, Portugal, The Netherlands, Germany, U.K., South Africa and Slovenia explore the methodological implications of their qualitative research on (or in) concrete military settings.

The book addresses a variety of topics and discusses the use of different research methods and techniques. Topics analyzed include the researcher's entry in the field, the organizational culture of the military, institutional frames for research, patterns of interaction and research relationships, the way hierarchy and discipline – two pillars of the military institution – affect the course of the research, gender-related dynamics, the "insider" and "stranger" status of the researcher, the publishing of research results, and ethical challenges of applied contract research. Among the variety of research methods and techniques stand out ethnographic and anthropological field research, direct and participant observation, in-depth interviewing and focus-group, document and content analyses and comparative research.

The guidelines for authors have been rather open regarding what to explore in terms of methodological issue topics and how to do it. However, all of them share a similar viewpoint regarding the need for reflexivity when conducting research and writing about experiences in the field. Acknowledgment of the complex set of social circumstances where research takes place in military contexts is a central concern to every chapter. The practice of reflexivity as a way to improve the quality of, and ultimately, accountability in the research process (Burawoy, 2003; Higate and Cameron, 2006), is the backbone of this book.

While the concept of reflexivity has been used extensively in qualitative approaches and its value has been largely accepted in most social scientific disciplines over recent decades, it has remained peripheral to the field of military studies. This marginality has been considered to result from a number of reasons: the dominance of a positivist epistemological paradigm and of applied psychological models; the marginality of the subdiscipline of military sociology within the wider field of sociology; and the impact on research of explicit military agendas oriented to making the armed forces more efficient and effective, thus promoting an engineering rather than an enlightenment model of social research (Kurtz, 1992; Higate and Cameron, 2006).

The starting point of the idea of reflexivity is that it is not possible for a social researcher to be detached from what he or she is observing. It is "a process that challenges the researcher to explicitly examine how his or her research agenda and assumptions, subject location(s), personal beliefs, and emotions enter into their research" (Hsiung, 2008). The researcher is conceptualized as an active participant in knowledge production rather than as a neutral bystander. This recognition involves a constant process of assessment by the researcher of his or her contribution and influence in shaping the research findings.

However, there are different interpretations and possible uses of reflexivity. One of them underlines the way the structural and existential position of the researcher determines what he/she can know about reality. The underlying argument is based on the assumption that position determines perception and cognition. In its most radical relativist version, this interpretation considers the "other" unreachable unless one becomes the other through experience; ultimately we would have to be like others to understand them, and consequently would have to reposition ourselves as researchers to be able to reach better understanding (Rosaldo, 2000). However, as Salzman notes in a review essay on the uses and limits of the concept of reflexivity in Anthropology, "the more general argument that we must be like others to understand them seems to doubt human capability of empathy, sympathy and imagination" (Salzman, 2002: 808). According to Salzman, the abstract reference to "position" without consideration of context and situation leads to the reification of that analytical category and to making the concept of reflexivity unhelpful (Salzman, 2002: 810, 812).

In this book, we try to avoid abstract descriptions of the effects of the researcher's position, by looking at them in specific social contexts. The various chapters provide an opportunity to examine positionality "in action rather than repose" (Salzman, 2002: 810) as well as the possibility of contrasting or mirroring experiences. In this sense, reflexivity can also be understood as involving the use of information from another to gain insights into oneself (Easterby-Smith and Malina, 1999).

More than putting together a few exercises of self-interrogation and self-report, we believe that a relevant methodological contribution to military studies might derive from a process of open debate and critique, issuing from the way the various authors expose their views, achievements and difficulties regarding the use of qualitative methodologies in military contexts. We understand reflexivity not only as the self-awareness of the researcher concerning his position in the field, but also the critical capability to make explicit that position and its impact on the research process.

In the first chapter, Celso Castro takes reflexivity to provide a broad overview of the research developed in Brazil by a dozen researchers who have conducted fieldwork in the armed forces during the past two decades. His chapter is based on personal accounts made by these researchers (assembled in the book "Antropologia dos militares" [*Anthropology of the Military*], Rio de Janeiro, Zahar, 2009, ed. Celso Castro and Piero Leirner) and focuses on some of the central issues of their field experiences, which the author considers important for the consolidation of an anthropological perspective on the study of the military. This review and discussion opens the stage for the other chapters of the book.

The following two chapters by Kirke and Langer and Pietsch explicitly address reflexivity and carry it out by focusing on the position of the researcher and the institutional frames for research. Charles Kirke describes the theoretical and methodological issues that emerged during a study on the organizational culture of the British Army at unit level between 1996 and 2002. Being a service member himself, the author reflects upon his "insider status" as a full member of the organization being studied. This led him to examine the nature of "insider-ness" and its variability in different contexts, which in turn led to his characterization of a typology of degrees of insider-ness. Although the challenges and issues in this chapter were worked out in the specific context of the British Army, they are also relevant to those who would investigate disciplined human groups of any kind, those who wish to use models in social science, and those who find themselves carrying out social science studies as an insider rather than as a stranger.

In the third chapter, Langer and Pietsch build on the question of the institutional frame for research, and discuss methodological and ethical questions of applied contract research for the armed forces. Drawing on a qualitative research on intercultural competence, carried out in 2010 at the German *Bundeswehr Institute of Social Sciences*, the authors address the particular challenges of applied contract research and present a critical and self-reflexive analysis of the very conditions of research from inside the armed forces and their methodological implications.

From a different angle, Ian Liebenberg takes us into a self-reflexive journey through the process of writing his own doctoral thesis on the link between the SATRC (South African Truth and Reconciliation Commission) and its potential effect on civilian control in a post-apartheid democracy, to discuss the uses and importance of auto-ethnography. Sharing the view that auto-ethnography adds value to socio-political understanding while liberating research findings from imposed objectivities or an omnipotent unchanging "universal truth," he argues in favor of auto-ethnography as a tool for military sociologists.

In his chapter "Side effects of the chain of command on anthropological research: the Brazilian Army," Piero Leirner further develops the topic of the duality insider/outsider in the framework of ethnographic research with the military. Beginning with a wide-ranging discussion of the idea of anthropologists working "with the military," he then seeks to situate the results of an ethnography carried out with the Brazilian Army through the symptoms and/or collateral effects that were visible both during and after his research. By taking up the relationship that was established, and also that which was not, it was possible to observe the centrality of the concepts of "friend" and "enemy" in the definition of a wide range of ties between the military world and the "outside world."

The problem of getting access to the military, when the researcher comes from this "outside world," is the focus of Alejandra Navarro's chapter. It explores the complex decision-making process that allowed the author to access two military institutions of the Argentinean Army between 2008 and 2010. Based on a study originally aimed at analyzing the social origins, patterns of sociability, spatial trajectories and motivations to enlist of three cohorts of the Argentinean Army, Navarro addresses the impact of a long negotiation process with key informants and potential gatekeepers on other components of the research design (specific objectives, selection of cases and construction of research tools). A central idea to this chapter is that access, understood as a dynamic and flexible process, should not be underevaluated since it might have considerable implications for the development of the research.

With a similar specific focus, here on the impact of gender on research in military settings, Carreiras and Alexandre discuss the variety of aspects through which gender might affect the course of the research. Drawing on the study of a Portuguese peacekeeping mission in Kosovo in 2009, they discuss the impact of the researcher's and the researched gender, the gendered interpretations of the researched, the gendered nature of the context and the gender focus of the research topic. The aim of this account is to illustrate how, in this specific case, researchers dealt with the gender dimension of the research process, including both the trade-offs involved in the negotiation of their roles, and forms of control required to acknowledge the impact of gender in the conduct of field research.

The following chapters by Vuga and Juvan and Deschaux-Beaume raise important questions regarding the suitability of qualitative methods for the study of the military, exploring the appropriateness and challenges of specific research designs, namely mixed-method and comparative approaches. Based on their experience of two decades of researching the Slovenian Armed Forces (SAF), Janja Vuga and Jelena Juvan examine the suitability of qualitative methodology for studying a military organization, underlining issues of reliability and validity. After reviewing a list of specific challenges raised by the very nature of the military as a research context, they sustain that triangulation of various methods, as well as (in some cases) of researchers, proves to be the most appropriate strategy for ensuring the reliability and validity of data.

In her chapter, Delphine Deschaux-Beaume explores the meaning of studying the military with a qualitative and comparative methodology. The chapter is based on the author's Ph.D. dissertation dealing with the genesis, practices and uses of the European Security and Defence Policy, with a focus on the comparison between France and Germany, both in the genesis and daily practices and representations of ESDP actors (military and diplomats). It raises three main and intertwined issues: first, qualitative inquiry and the confidentiality of military missions; issues of comparability in qualitative approaches in the military field, especially regarding interviews; and finally, reflexivity regarding the position of the researcher vis-à-vis military officers.

Two other chapters, by Saïd Haddad and Dirk Kruijt, explore concrete research techniques and the way these were used in different research experiences: focus group and interviews. Haddad's chapter is based on research on the French military's perceptions of a multinational operation. It discusses the specificity of the focused or group interview technique in a military context, and deals with a variety of factors which affect research dynamics and results: the position of the researcher, including his or her autonomy in the selection of the interviewed and his or her responsibility in the material organization of interviews; the legitimacy of the groups and data reliability; the real purpose of the interview; taking account of the nonverbal elements which shape an interview; and the political and social context where the research takes place.

Dirk Kruijt's contribution is a colorful and detailed account of the author´s research based on in-depth interviews with higher echelons of the Latin American military and guerrilla leadership since the mid-1980s. Using an intense and lively narrative style, the author describes the development of a research style where conversational intimacy, shared confidence, sensitiveness about the military ethos and culture, esprit de corps and a culture of silence vis-à-vis "civilians" are paramount. In Kruijt's proposal, this style of interviewing is the fundamental instrument for assessing vital non-official contemporary history: counterinsurgency operations, peace negotiations, agreements between the military and "civilian" politicians, pre-coup bargaining between economic and political elites and posterior military governments.

The idea for the book originally came from the session on "Methodological problems in the study of the military," organized by the International Sociological Association's Research Committee 01 "Armed Forces and Conflict Resolution," during the XVI World Congress of Sociology that took place in Gothenburg, Sweden, in July 2010. The interest shown on the occasion of the session's call for papers, and the discussions during its course, convinced us that it would be worth engaging a wider public in the debate in order to encourage more discussions on the use of research methods within the empirical context of the military institution. This is the main purpose and proposal of the book.

References

Burawoy, Michael (2003) "Revisits: An Outline of a Theory of Reflexive Ethnography," *American Sociological Review*, 68(5): 645–678.

Easterby-Smith, Mark and Danusia Malina (1999) "Cross-Cultural Collaborative Research: Toward Reflexivity," *The Academy of Management Journal*, 42(1): 76–86.

Higate, Paul and Ailsa Cameron (2006) "Reflexivity and Researching the Military," *Armed Forces and Society*, 32(2): 219–233.

Hsiung, Ping-Chun (2008) "Teaching Reflexivity in Qualitative Interviewing," *Sociology*, 36(3): 211–226.

Kurtz, Lester R. (1992) "War and Peace in the Sociological Agenda," in Halliday, Terence C. and Morris Janowitz (eds.), *Sociology and its Publics*, Chicago, University of Chicago Press.

Rosaldo, Renato (2000) "Grief and a Headhunter's Rage," in McGee, Jon R. and R.L. Warms (eds.) *Anthropological Theory*, 2nd edition, Mountain View, CA, Mayfield Publishing, pp. 521–535.

Salzman, Philip Carl (2002) "On Reflexivity," *American Anthropologist*, 104(3): 805–813.

1 Anthropological methods and the study of the military

The Brazilian experience[1]

Celso Castro

There are still very few anthropological studies on the military institution. Among the authors or books considered classics of the discipline, present on the syllabus of the mandatory courses, none of them approach the professional military as a subject of investigation. Only in the last decades have anthropologists begun, although in small numbers, to study the military.

In Brazil, during the past two decades, about ten researchers have done fieldwork in the Armed Forces using participating observation, a classical method of ethnographic research since its relevance was established in the beginning of the twentieth century by "founding-fathers" of modern anthropology such as Franz Boas and Bronislaw Malinowski. This chapter is based on personal accounts made by these researchers (assembled in the book "Antropologia dos militares" [*Anthropology of the Military*], Castro and Leirner [2009]) and focuses on some of the central issues of their experiences: the researcher's entry in the field; patterns of interaction with the "natives"; the way hierarchy and discipline – two pillars of the military institution – affect the course of the research; gender-related dynamics; and issues related to the publishing of the researcher's results.

This research was conducted in the last two decades, a period of renewed interest in the comprehension of the "military world" in Brazil. Before this, the majority of the work produced in the fields of Social Science and History focused on military interventions in politics (especially through insurrectional movements or *coups d'état*) or on the transition from the 21-year-long military regime the country lived under between 1964 and 1985 (with emphasis on the analysis of the military subordination to civil power).

The reason for the pre-eminence of the approaches centred on the political dimension is easily explained. Since the establishment of the Republic in Brazil by a military coup in 1889, the military was, throughout a century, a fundamental actor in Brazilian history, promoting several other coups and interventions, although power was quickly restored to the civilians. The exception to this pattern was the direct exercise of political power between 1964 and 1985, when Brazil was successively governed by five general-presidents. Beginning in 1985, with the transition to a civil government, the process of re-democratization and the birth of the "New Republic," the military gradually lost political importance in Brazil. It is important to point out that, in the last quarter of the century, it never posed a significant threat to democracy. The experience of the military regime, however, left marks on Brazilian society, including upon the intellectuals, who were strongly affected by authoritarian acts.[2]

The anthropological approach of the Brazilian military began after the end of the military regime and recognizes the importance of the work produced, for the most part, by political scientists in the 1970s and 1980s.[3] It gives, however, centrality to new research themes: daily life in the casern; the process of professional socialization; the construction of military identity; family life in military communities and buildings; and the relationship between the military and their ethnographers. As such, it seeks to contribute to a denser perception of the military world, generating a better understanding of life in the barracks, as well as to broaden the issues faced by anthropology in the study of a central state institution in our own society.

Two worlds: the military and the civilians

I was a pioneer in the anthropological study of the Brazilian military in conducting fieldwork at the Military Academy of Agulhas Negras (AMAN) between 1987 and 1988 and defending, in 1989, my master's thesis entitled "O Espírito Militar" (*The Military Spirit*), published as a book in 1990. The AMAN is the only undergraduate school that trains career officers of the Brazilian Army, through a four-year course under a boarding school regime. The focus of my research was the process of construction of the military identity, as experienced by the cadets of the AMAN throughout their professional socialization. I sought to overcome an "external" view regarding the military, which is inclined to contemplate them through an exotic and ethnocentric point of view, and to obtain an "internal" view of their social world, so as to understand how military identity is constructed and how its world view is structured.

It is important to stress that the use of the words "internal" and "external" does not refer to any supposedly reductionist view on my part, an approach that would "isolate" the military institution from the society of which it is a part. On the contrary, it is the main result of the effort to understand fundamental "native categories" of the military, based on symbolic distinctions produced by the military between "inside/outside" and "military world/environment" (*mundo/meio militar*) versus "civil world/environment" (*mundo/meio civil*). These categories are not descriptive terms: they are structural to the military world view. To join the Armed Forces means, above all, *not to be a civilian.* A point of consensus among the researchers assembled in this book is the perception that the opposition between civilians and the military is integral to the military identity.

One pole of this relation – the military – is hierarchically superior in terms of values to the other, the civilian – or *paisano*, as they say among themselves. This superiority doesn't have an individual origin, but a collective one. The military feel like they are part of a military "world" or "environment" that is superior to the civilian "world" or "environment" – the world of the *paisanos*: they represent themselves as more organized, more honest and more patriotic.

It is important to emphasize that these categories are updated in interactions that are not a-historical. That is, they are inserted into a "field of possibilities" that is historically and culturally marked and that modifies with time, although more slowly if compared to other institutions. It should be noted that the very result of the research that civil investigators conduct on the institution can influence these interactions. What is most important, in the theoretical plane, is not to lose sight of the fact that categories like "civilian" and "military" pertain to a logic that is simultaneously *situational* (that becomes effective in specific contexts, in which these categories "come into action" and can be reaffirmed, questioned, modified or transformed) and *relational* (they only exist in opposition to other categories – such as "civilians," "enemies," etc.)

The contrasting and permanently reaffirmed relation between "in here" and "out there," with due perception of their differences, is the fundamental aspect of the process of social construction of the military identity to which the cadets of the AMAN are submitted. The initial period, in particular, deceivingly called "adaptation," is full of examples of a sudden and sharp symbolic rupture with the exterior world. Since the very first moment, mechanisms that Ervin Goffman (1961) called "mortification of the ego" are put into action, which remove the individual's previous "identity kit."

The entry into the "military family" – a recurrent theme throughout our book – is obtained by the cadets through a process of professional socialization under a boarding school regime, with few and rigidly controlled hours of leave. This reduces the weight and intensity of their family bonds – the feeling of longing for their family of origin is an important part of the professional socialization to which the cadets are submitted. That is, the "secondary socialization" that they undergo in the Military Academy acquires an extreme form of alternation (Berger and Luckmann) that marks the entrance into a new family – the "military family," a process that attempts to recreate the emotional weight experienced during primary socialization. When entering the military academy, the youngster undergoes a process of construction of military identity that presupposes and requires the deconstruction of his previous "civilian" identity and the construction of a military "self." Even when he transits through the so-called "civil world," the military does not cease to be a military – at most, he can dress in civilian clothes.

Throughout the military career, there is also a great concentration of interaction inside the same "social circle," to use one of Georg Simmel's images. Due to this, the "military world" becomes more differentiated, while the individuality of its members becomes more undifferentiated. In military life, in addition to the workplace, the place of residence, leisure and schooling is also shared, in great measure. This characteristic is extended to spouses and sons, embracing the whole "military family." The endogenous social interaction is formally stimulated, through co-fraternization events organized by the institution, as well as informally, through social encounters organized by military colleagues. The role of the wives (and to a certain extent, the sons) is fundamental. An informal, yet obvious, reproduction of the husbands' hierarchy takes place among the military wives.

It is important, however, to denaturalize our own conception that there are in fact "civilians" or a "civilian world/environment" – a common taken-for-granted view not only among the military, but also among many researchers who study the so-called "Civil–Military Relations." The civilian is a military invention. I am only a civilian in relation to the military and when I am classified by them as such. If I had to make a list of the main elements that define my identity, "civilian" wouldn't appear among them. For any military, however, being a military appears among the first attributes, if not the very first. This is due to the fact that they are part of an institution that I prefer to call "totalizing," to distinguish from Goffman's idea of "total institution," many times applied to the military world (inadequately, in my opinion).[4] By making this change in terminology – from "total" to "totalizing" – my intention is to better characterize the basic and totalizing experience towards the military identity that embodies and establishes the differential characteristics between military and *paisanos*: the pre-eminence of collectivity over individuals. The result is the representation of the military career as a "total career" in a coherent world, full of meaning and where people "have bonds" among themselves.

Perhaps a differential characteristic of the anthropologist who makes use of participating observation in a military institution in relation to colleagues of other disciplines that do not make use of this research method is precisely the personal experience of feeling like a civilian, or in the depreciative native version, feeling like a *paisano* – something that is not usually part of our social identity. Not only does it constitute an intellectual experience, but also an existential one, in the widest sense. It involves, for example, the perception of a different corporality – something fundamental to the military.

Towards an anthropology of the military

As I said, *Antropologia dos militares* shows the result of ten research projects that applied, in different degrees, the method of fieldwork with participating observation to study the military institution. The majority of these authors participate in the same academic network, with shared experiences and references to a common bibliography. Not all of them, however, identify themselves professionally as anthropologists. Some are sociologists, political scientists or historians. Nevertheless, all of them had, to a lesser or greater degree, their research experiences marked by the anthropological production that is available, either by academic or personal relationships with anthropologists who have researched the topic. Above all, at some moment, they all went to the field and lived with the military "in the flesh," observing or participating in their daily activities, and not just resorting to archival data and interviews. Besides reading documents and texts and listening to military personnel interviews, they also had the opportunity to observe aspects of the military everyday life in action.

It's important to point out some limits of the research assembled in the book. There is a much larger concentration of studies on officers than on soldiers and recruits performing mandatory military service; more studies on the Army than on the Navy and the Air Force; on specific moments of the military career rather than on its full course, from the initial training until retirement. In addition to this, there are still few comparative studies between different generations of military and with the experience of other countries – with the exception, in this case, of the chapter written by Máximo Badaró, which presents his research on the Argentinean military.

I am not sure to what extent this collective experience can be extended to other countries. The military institution undoubtedly possesses a high degree of cosmopolitanism, through which military from different countries share many elements that are common to their profession. On the other hand, there is also the undoubted influence of different national realities. Thus, I hope that the general observations that follow may, if not help to elucidate the issues experienced by researchers in other countries, serve as comparative material.

I will attempt to summarize and condense a collective research experience, although I risk misinterpreting or not emphasizing differences in the experiences of each author or, especially, giving more centrality to my own experience. However, I believe that the topics I will present are recognized by all, to a lesser or greater degree, as constitutive of their individual research experience.

Fieldwork in the "military family"

Nearly all of the researchers assembled in the book had little or no contact with the military institution prior to conducting their research. The only exception is my personal "son of a military officer" condition. My father was an Army officer, and because of this I frequently lived, throughout my childhood and adolescence, in military communities and buildings and studied in military schools for two years. This condition imposed itself on me as a necessary exercise of anthropological self-reflection.

During the research at the military academy, a conscious strategy used both by the military and myself, in order to reduce the symbolic distance between us, was to always mark my belonging to the military family. The "military family," more than a biological bond, is an extremely important native category and reflects some fundamental values of the military world. The family is seen as an extension of the barracks, which, at the same time, is also reached by the family. To a considerable degree, the hierarchical position of the military person extends to his family in interactions with other military families. The apparently informal sociability, during parties, for example, separates the "cliques" by hierarchical circles – but also by gender. One exception refers to the military women, a situation that is still relatively new in the Brazilian Army.

In the case of my research, I had already indicated my "military son" condition in the letter I wrote requesting authorization to research at the AMAN. The military always conveyed the information that I was the "son of a comrade" (*filho de um companheiro nosso*). In addition, we also accentuated the fact that I had studied in military schools. The condition of being the "son of one of our comrades" turned me into a potential "friend" of the Army. That was significant. A common experience for the researchers assembled in the book is the fact that they were always subject to being classified as "friends" or "enemies" by the military, which was fundamental for the success (or failure) of their research. Even if political or ideological divergences presented themselves, it was necessary to create some level of "trust." Perhaps this is a consequence (and a requirement) of the ideal "combat" situation – structural to the military cosmology – in which there cannot be any doubt regarding the classification of the person with whom one interacts.

The condition of being the son of a military officer distanced me from the stereotype of a civilian who is viscerally hostile towards the military, as the military perceived, not without reason, a substantial part of the Brazilian media and academic world at the time. Differently from other civilian-academics, I did not feel a prior emotional repulsion towards the military, despite my disapproval of the role they played in politics and of some aspects of the military way of life. My "son of a military" condition, however, was in no way a total warranty of trustworthiness. On the other hand, the fact that I was a civilian researcher, a "sociologist," as I was sometimes characterized (which, to the military at that time, was similar to being a "socialist") was something that was always present and reaffirmed during the research, placing me beyond, or at least at the margins of the military world. This hybrid condition – civilian researcher and military son – granted me powers, but also invoked symbolic risks. My belonging to the "military family," less than a biological fact, was a social condition in a permanent process of negotiation that had to be constantly reaffirmed. Raymond Firth wrote that sometimes "the anthropologist as an observer is a moving point in a flow of activity." Nothing could better describe this permanent transformation of my identity in the field.

Access and control

It was of fundamental importance for all the researchers in the book to have institutionally approved access: it was always necessary to have some kind of official authorization, mediated by someone in a prominent position in the military hierarchy, in order for the research to proceed. Once authorization was obtained, however, the degree of control that was sought over the research varied hugely among the experiences assembled in this book. In some cases, the researcher had a lot of freedom: knowing he "was authorized" by superiors was enough; in other cases, the daily routine of the research had to be thoroughly established, as well as closely followed by an officer especially designated to accompany the researcher and mediate the contact with his "natives."

I believe that although the fact of doing fieldwork with participating observation in the casern can be feared by the military in certain situations, it can also be valued in others – the fact that the researcher is not, in the eyes of the military, bound to "external" prejudices and that he is willing to be among the military personnel and approach their real daily life. I think these characteristics make anthropological research and its results – what we say and the texts we write – more difficult to accommodate in the "friend/enemy" scheme than in the case of colleagues of other disciplines that study the military – more difficult, for example, than a political scientist or a historian that observes the casern from an "external" point of view and discusses, primarily, the moments of coups or military intervention in politics.

Often, in the case of the researchers assembled in this book, it was a moving experience to be classified as a "friend" of the military through *rites of passage*, which make the researcher become trustworthy. At these moments, questions of why he does not enter the career are common. In the case of the female researchers, jokes about how they could end up marrying a soldier were recurrent. Beyond the more personal dimension of social interaction, we can perceive how the *familial metaphor* (in which, in its traditional version, the patriarch has a position of moral authority and hierarchical superiority) possesses political relevance in the relation of the military with the civilians: it suffices to observe the recurrence of images of the Army as guardian of the nation, of the casern as a continuation of the family (through mandatory military service, etc.).

Another common aspect of the research of the authors assembled in the book is the great interest that the military manifested regarding the result of the research, what we would say about them, if we would say "good" or "bad" things about them – which is obviously related to the classification of potential "friends" or "enemies."

My final point relates to the fact that these research experiences in the military institution fit into a wider experience of an *anthropology of the elites*. To do fieldwork in someone's own society and among socially privileged groups – even if in more symbolic than material terms – puts us face to face with some peculiarities. The vast majority of anthropology has been and continues to be done with groups in some way socially subordinate in relation to the anthropologist, or that do not master his academic language. The research with elites, however, could invert the direction of this relation of domination/subordination. At many times during the research experiences assembled in the book, it became evident that some of our "natives" felt that their intellectual, social or moral position was superior to that of the researchers. On the other hand, the reverse experience of dealing with the prejudices of the Brazilian academic world was also common. To some extent, the study of military was not considered "noble" in the Social Sciences field in Brazil; on the other hand, to study them in a society such as the Brazilian one that suffered with the experience of a military regime, was seen as something potentially polluting.

Notes

1 A preliminary version of this chapter was presented at ISA 2010 Conference, in Gothenburg (RC01).

2 On this subject, see Bethell and Castro (2008).

3 See, in particular, the works of Barros (1978), Carvalho (2005), Coelho (1976) and Costa (1984). The common point between these researchers is the problematization of the perspective that dilutes the specificity of the military institution either by linking it to a theory of class struggle, or when the political behavior of the military is explained by its presumed "middle class" social origin. According to the authors mentioned above, the importance of social origin in defining the political role of the military, on the contrary, is only marginal, which implies the recognition of the institution's *relative* autonomy.

4 Goffman includes *en passant* barracks and military academies as examples of total institutions, although he uses prisons and mental asylums as the basic references for the construction of his ideal type of "total institution." I believe, however, that more is lost than gained by classifying military academies in this way, for there they show great divergences with Goffman's model, despite the many formal similarities: (1) A rigid social division between the directive staff and "inmates" is inexistent. The chain of military command is not of the same nature. Although there is an unbridgeable barrier between officers and soldiers, there are strong mechanisms of conviviality and social mobility based on individual merit within the officers' stratum (including the cadets). (2) Goffman makes it clear that total institutions do not seek "cultural victory" over the internee, but instead the maintenance of a tension between his domestic world and the institutional world, in order to use this persisting tension as "a strategic force in the control of men." In a military academy, on the contrary, a "cultural victory" is sought, and not the creation of a "persisting tension": the academy is clearly seen as a place of passage, a stage to be overcome. (3) Goffman's model mainly refers to establishments of compulsory participation. In a military academy, however, only those who wish to remain do so.

References

Barros, A. de S. C. (1978) "The Brazilian Military: Professional Socialization, Political Performance and State Building," PhD dissertation in Political Science, Chicago: University of Chicago.

Berger, P. and Luckmann, T. (1966) *The Social Construction of Reality: A Treatise in the Sociology of Knowledge*, Garden City, NY: Anchor Books.

Bethell, L. and Castro, C. (2008) "Politics in Brazil under Military Rule, 1964–1985," *The Cambridge History of Latin America – Volume IX, Brazil since 1930*: 165–230, Cambridge: Cambridge University Press.

Carvalho, J. M. (2005) *Forças Armadas e Política no Brasil*, Rio de Janeiro: Zahar.

Castro, C. (1990) *O Espírito Militar: um estudo de antropologia social na Academia Militar das Agulhas Negras*, Rio de Janeiro: Zahar, 2nd ed. rev. and updated (2004) as *O espírito militar: um antropólogo na caserna.*

Castro, C. and Leirner, P. (eds.) (2009) *Anthropologia dos Militares: reflexões sobre pesquisas de campo*, Rio de Janeiro: FGV.

Coelho, E. C. (1976) *Em Busca de identidade: O Exército e a Política no Rio de Janeiro.* Rio de Janeiro: Editora Forense Universitária.

Costa, V. M. R. (1984) *Com rancor e com afeto: Rebeliões militares na década de trinta*, Rio de Janeiro: CPDOC/FGV.

Goffman, E. (1961) "On the Characteristics of Total Institutions," in *Asylums: Essays on the Social Situation of Mental Patients and Other Inmates*, New York: Doubleday.

2 Insider anthropology

Theoretical and empirical issues for the researcher

Charles Kirke

Introduction

The Battery[1] Warrant Officers and Sergeants[2] were entertaining the Officers and their ladies to dinner in the Sergeants' Mess.[3] The new Battery Commander ('the BC') was the guest of honour, sitting next to the Battery Sergeant-Major[4] who was presiding. The tables gleamed in the light of the candles, which also gently lit the Battery's silver collection. As the mess waiters came and went with food and wine a warm glow seemed to come over the room and its occupants. Oddly, though, as the BC casually observed, the level of wine in the glass of one of the troop sergeant-majors, Sergeant-Major O'Donnell,[5] never seemed to change. Were the waiters constantly refilling his glass or was he just not drinking?

When the meal was over then came the toasts ('The Queen', and 'The Battery'), after which hosts and guests dispersed to the anteroom to enjoy the rest of the evening drinking and talking with each other, or dancing in the disco in the mess bar. At this stage Sergeant-Major O'Donnell came to talk to the BC. 'Sir', he said, his voice slightly slurred, 'There's something I think you ought to know.... It's the way we're treating Sergeant Mallerby. I've known him for a long time and I don't think he's being given a fair chance.'

Over the next few minutes the sergeant-major told the BC many things that he thought he needed to know about Sergeant Mallerby, and he was probably right. But for all his slightly tipsy demeanour, his breath did not smell of alcohol at all.

This incident is rich with cultural fuel for anthropological research. An anthropologist who was present could look at the distinctive groupings (Sergeants and Warrant Officers, and the Officers), they could look at the special nature of the place in which these events were taking place (the Sergeants' and Warrant Officers' territory, which no officer or junior soldier ever entered unless on duty or by invitation), they could look at the time of day (evening), or the material culture represented by the artefacts on the table and the way that they were lit, they could observe the way the toasts were made, they could analyse the conversation, they could observe the behaviour of the women (were they 'in' the Battery, were they marginal to the Battery?) they could listen to and record the various terms of address, and they could still come out with an incomplete analysis of the totality.

If we look at Sergeant-Major O'Donnell's behaviour in particular, we see a man who wanted to appear slightly drunk but was not. This tells the anthropologist not only that the context was such that drinking alcohol was perfectly acceptable, but also that to have this sort of conversation with the Battery Commander it is best to be thought to be slightly drunk. That could lead to another line of inquiry about informal relationships across differences in rank and 'acceptable' and 'unacceptable' topics of conversation within those relationships, and how those categories are altered with the consumption of alcohol.

But what if it is the Battery Commander who is the anthropologist?

This chapter examines the situation in which its author found himself in researching the organizational culture of British soldiers[6] in combat arms units[7] at regimental duty[8] as a social anthropological project conducted from within the organization. The project was started in 1974 as an informal and occasional personal study following a first degree in Archaeology and Anthropology, and it progressed in a second, more formal, phase through a year's research at Cambridge University in 1993/1994, and a subsequent PhD at Cranfield University in the organizational culture of the British Army at unit level[9] between 1996 and 2002 (Kirke 2002). It has led to a number of further publications, the most comprehensive of which is Kirke (2009).

Although this research was focused on a particular set of human groups (units of the British Army) the issues explored here appear relevant to the general context of anthropology carried out by a member of the human group being researched. By its nature, therefore, this chapter is autobiographical and reflexive – the distillation of one person's experience in a particular set of situations, generalized for the wider context. And because of its autobiographical origins, the text that follows will use first person pronouns rather than the more academically strict third person.

This chapter is also, of course, retrospective. As Lieven warns us in another context about autobiographical work, 'For most people, life unfolds in a patternless way bewildering to its subject. By contrast, the autobiography is typically written by someone looking to find a sense and pattern...' (1999: 107). The issues that this chapter explores are therefore constructed from experience modified with hindsight: many of them emerged chaotically in the day to day process of carrying out the research rather than as the structured wholes in which they are presented here.

The first issue to be considered in this chapter is the nature of 'insider anthropology', where the researcher has something in common with the researched before the research begins. The second is the research aspect of 'stranger value' and its uneasy relationship with anthropology from within. Third, this chapter addresses the practical issues involved in the author's research within the British Army and the generalized conclusions that may be drawn from it.

Insider anthropology

Cerroni-Long's edited NAPA Bulletin *Insider Anthropology* (1995) contains a number of articles written by individuals who had had experience of investigating the culture of human groups with which they had something in common before they started. All of them, as implied by the title of the bulletin, could be called exercises in 'insider anthropology'. However, an interesting spread of 'insider-ness' emerges from the different texts in the bulletin. For example, Walter Goldschmidt, a native of California, studied 'California rural communities in the early forties' (1995: 17) and in the course of his chapter he poses a question as to whether or not his research area ('Wasco') was really representative of his own culture (1995: 18). In contrast, there is no doubt about the 'insider-ness' of Alexandra Jaffe when she attempted to produce an ethnography of that part of the US Army in which she served, a full 'insider' experience (1995). Another article in the bulletin, by Edward Liebow, concerns a study carried out among agribusiness people to whom he did not belong, but who were fellow-Americans (1995: 22).

'Insider Anthropology', therefore, can mean different things even within the same academic publication. This implies that the term can be broken down into more meaningful sub-categories to separate the various degrees of 'insider-ness'. Further reading in the literature shows that there is a range of insider-ness: at one extreme there are studies of bounded exotic, or 'other', groups that live within the country of the researcher, an example of which is Okely's study of Gypsies in England (1983a, 1983b). In many ways researchers in these contexts are hardly 'insiders' at all: even though they and the group may share a common language and live in the same country they scarcely share a common culture in any significant sense. Researchers in these fields are barely distinguishable from the traditional anthropologist who travels abroad to explore 'the other', apart from some practical advantages in terms of the availability of home resources, the likely sharing of a common language (but probably not a common dialect) and reductions in time and money spent in travelling. This type of study, referred to in this chapter as '*anthropology of the "other" at home*', requires much the same skill set as conventional anthropology in exotic settings.

At the other extreme come the genuine 'insider' studies where full members of a group attempt to research the culture of that group, called here '*anthropology from within*', whose peculiar advantages and challenges have been thoughtfully explored by Labaree (2002). In summary, he categorizes the advantages of this type of research

> into four broad values: the value of shared experiences; the value of greater access; the value of cultural interpenetration; and the value of deeper understanding and clarity of thought for the researcher. Each

> of these advantages also has concurrent challenges that the insider participant observer must negotiate and come to terms with.
>
> (2002: 103)

These challenges include ethical issues surrounding disclosure of what the members of the group say and do, and the achievement of 'stranger value' – the ability to look at people in a way that is not flavoured by the culture which they and the researcher share. This latter in particular could lead to difficulties in maintaining objectivity and accuracy and the preservation of sharpness of observation in what to the researcher is a mundane environment, and in the network of power relationships within the group (2002: 106–109). Such *anthropology from within* is exemplified by Young's study of the Northumbria Police Force (1991), Lofgren's study of 'Swedishness' (1987), Collins's of Quakerism (1998), and Fox's *Watching the English* (2004).

A middle position between these extremes is represented by researchers who are familiar with the human groups concerned, but are not full members, for example Forsythe's work on employees in artificial intelligence laboratories (1992, 2001), a field in which her parents were deeply immersed but to which she did not belong. Another example is Irwin's research on a Canadian infantry company, carried out when she was a reservist from a different part of the Canadian Armed Forces (Irwin 2002). Such work is called here '*anthropology of the familiar*'. This position shares many of the advantages and challenges of *anthropology from within*, but access is not as easy and it is less likely that the researcher will share the deep assumptions and attitudes of the group being studied. They may, however, find themselves more personally engaged with an aspect of the field than in conventional anthropology of the 'other'. In the case of Forsythe, for example, she knew a great deal about the work that was undertaken in the laboratories in which she carried out her research and about the ethos of the workers. However, she was deeply unsympathetic to the general attitudes of those workers towards the eventual users of the systems which they were developing, and this led her to appear outspoken in condemnation of them (Fleck 1994). Such apparently emotional engagement is unusual in anthropological studies.

This typology of different degrees of insider-ness within the overall category of 'insider anthropology' can be demonstrated when considering my study of the British Army. At the holistic level, the research fell into the *anthropology from within* category as I was a full member of the institution which I was researching. However, when the research is examined in more detail, different terms apply. Although I was a full member of the Army, I was not a full member of any of the particular military groups which I researched after 1986, as I was not at regimental duty studying my own unit. In the small scale, therefore, after 1986 I was more in the position of an *anthropologist of the familiar*, visiting several different individual units rather than concentrating (as hitherto) on the unit to which I actually belonged. This distinction is much more than merely semantic: when researching 'from within' I was invisible to the organization, simply another member pursuing a personal interest and blending in with the other 400-odd members in my particular place within the organization; when researching 'the familiar' I remained outside the unit organization, recognized as a member of the wider institution of the Army to which we all belonged, but a guest to the unit, an outsider with no access by right to any part of it.

How do other ethnographic works on units of the British Army appear on this scale? There are very few, but they show an interesting spread. The earliest attempt at ethnography is John Hockey's *Squaddies* (1986). Hockey had seen some service with the British Army (unspecified in his book) before his research (which comprised three, one-month, periods of immersion among private soldiers) so he had some domain knowledge and an awareness of how to communicate with his research group. He was not, however, a member of the British Army and so his research must count as *anthropology of the familiar.* Paul Killworth's PhD thesis on culture and power at unit level in the Army (1997) is *anthropology of the 'other' at home* because he knew nothing of British military culture before he began to research it. The same applies to Anthony King's exploration of group cohesion in the Royal Marines (2006) as he has never been a member of a military unit. As we saw above, my research has to be classified separately as *anthropology from within* when I was researching my own unit but *anthropology of the familiar* when I visited other units. These classifications indicate, for example, that when I was researching from within I had greater access to more areas of the field, and the advantages of shared formative experiences and deep understanding of what the others were experiencing. On the other hand, I found anthropological detachment difficult and had to generate particular stratagems to create it.

Stranger value for the anthropologist from within

A potentially serious difficulty for the insider anthropologist is that they are likely to be very closely involved in the daily life of the group being researched, and in danger of being unable to enjoy the detachment or 'stranger value' or 'otherness' which is a priority in conventional Anthropology (for example, Beattie 1966). Indeed, this was the insurmountable difficulty encountered by Jaffe in her attempt to carry out fieldwork on the US Army while a member (Jaffe 1995), and she had to give up her intended study because she found herself too closely identified with the organization. The conventional assumption is that a 'stranger' is in a position to observe without their observations being conditioned by belonging to the culture that is being observed. Thus they will notice the significance of mundane features which would otherwise be taken for granted by a member of the group being observed. Furthermore, their conclusions are more likely to be free from contamination or distortion by elements in the culture of interest. Such a 'stranger' should be able to place an interpretation on the data that it is neutral and, it is hoped, scientifically detached. Similarly, they should find it more possible than an insider to retain distance from external or internal groupings with special interests in the group being studied (Beattie 1966: 87; Fox 2004: 3, etc.).

Although the concept of 'stranger value' has come under attack as unrealistic – insofar as everybody has a point of view and this will colour their perception (for example, Clifford and Marcus 1986; Burawoy *et al.* 1991) – it nevertheless remains an ideal to be striven for. Some device must therefore be used by the insider researcher to create some form of artificial distance from the group as far as it is possible. Peter Collins gives us an interesting description of such a device in his research on a Quaker group of which he was a member. In his attempt to occupy a position pitched as exactly between the' insider/outsider' as the slash between the two words, he divided his work into three parts: a presentation of a conversation, a reflection on the conversation as an insider ('Simon') and interpretation as 'Peter' the outsider anthropologist (Collins 2002: 77). In my own case I made a similar effort self-consciously to create an attitude of detachment by effort of will, asking myself 'What would the anthropologist notice?' when confronted with data. This involved such techniques as striving to view information 'as if' it had been novel and alien, and continuously to look for differences between the 'rules of the game' as described by the soldiers and the daily practice revealed by my observations and their descriptions of incidents, anecdotes and situations.

There were, however, three advantages in my study which arose from the lack of 'stranger value' which are its equal and opposite and, I believe, are generalizeable to *anthropology from within* as a whole. First, both myself and my informants knew that we shared our organizational culture on a deep level and so it would be difficult for them to mislead me about aspects of their lives should they wish to do so – inconsistencies or countercultural aspects of what they said would be detectable and questioned. Second, I discovered that, because I shared the basic perceptions and assumptions of my informants and those whom I observed, the conclusions derived during the research (however ethnocentric) really did reflect the attitudes, assumptions and expectations arising from the culture being explored, and were in harmony with the social milieu as experienced by both researcher and researched. I represented these conclusions in a set of three social models and, once they had become relatively mature, every time I explained it to soldiers they accepted and embraced them as a fitting vehicle by which to describe, analyse and explain life in a unit. Third, while it is true that I was less likely than a 'stranger' to be able easily to set my observations in an external and detached framework (though I attempted to do so) I was much less likely than an outsider to import misconceptions into my analysis. As Scheurich (1997: 1) points out,

> [Researchers from outside the culture they are studying] are unknowingly enacting or being enacted by 'deep' civilizational or cultural biases, biases that are damaging to other cultures and to other people who are unable to make us hear them because they do not 'speak' in our cultural 'languages'.

Although lack of 'stranger value', therefore, is a source of potential weakness for the insider anthropologist I found that there can be significant compensating strengths.

These findings confirm and expand on the views set out by Labaree cited above (2002): for example, maintaining objectivity and accuracy when I was a fellow-participant in the culture I was studying was a constant concern; on the other hand our shared experience enabled me and the soldiers to understand each other and allowed me to be confident that I could detect any inconsistencies in their statements or attempts at deception or half-truths; and access to the research population was comparatively simple, even outside my own unit. In addition, I found that I was able, in constructing the models, to produce material that was equally well understood by the research population as by me.

Practical issues for the insider anthropologist

This section describes the practical issues that became apparent during the study. The first was the matter of access. In the conventional paradigm of Anthropology, the anthropologist lives with the target group of their research for long periods, sharing their lives, and becomes as much as possible a member of the group. In such conditions, typically in a foreign land and among people whose language the researcher has to learn, gaining access to a human group is likely to present considerable difficulties. Apart from the obstacle of language, the researcher will almost certainly require some form of official/local political approval and sponsorship. They will also require significant funding for travel and subsistence and to provide for the special research expenses needed in their chosen area. Once these obstacles have been overcome, there is the challenge of gaining the trust of the group being researched to the extent that they will be open about their lives and customs.

For the *anthropologist from within*, however, access is simple, and the process of becoming a member of the group is eliminated because the insider by definition already has membership. Blending into the group and its cultural background is not a problem. Similarly, no new language has to be learned – not even the special jargon of the group. Whereas some form of formal approval may be required to carry out the study in the first place, the insider has the opportunity to carry on the research as part of their normal life. And if the research is done in the work place, the task is likely to come with a salary, which reduces or possibly eliminates the need for research funding.

In my case, the concomitant of easy access during the early phase was a lack of time to conduct research in parallel with a demanding series of jobs within the units to which I belonged, and it seems likely that this will often be the case for the *anthropologist from within*. I addressed this difficulty by writing field notes as and when it was possible, and allowing the research activity to stretch over a considerable period (1974 to 1986) – to research extensively rather than intensively. In the second, more intense, phase (1993 to 2002), a period identified above as *anthropology of the familiar,* I gained short-term access to units by arranging for visits in a culturally sensitive way, exploiting my insider knowledge of what would constitute an acceptable approach. Time pressures were less extreme in this period because I was able to carry out the research as an integral part of my formally allocated work. Thus, as I lost my status as an *anthropologist from within* I found myself, ironically, with more time to devote to the research.

The second practical issue was in the gathering of data. In the early part of the study I had been able to observe soldiers' behaviour and material culture and to talk with them as part of my normal daily life, recording the observations in field notes. However, during the second phase, apart from observation it became necessary to undertake semi-structured interviews because I no longer lived and worked with members of the research population. These interviews provided information that would not have been available to me through observation alone (such as the practices and processes in groups of which I could never be a member), and it also allowed me to discuss some of the issues that remain latent in everyday life.

The differences between the two parts of the study highlighted the significance of the position of the insider anthropologist in the group which they are researching. It became clear that different positions within a group have different perspectives on the group and its members, the subgroupings and the cultural aspects. As Sanger puts it, 'information is that which an individual perceives as significant' (1996: 6), and individuals in different positions will perceive the same elements in different ways.

This issue was particularly stark in my case in the second phase. An outsider to the Army (whatever their viewpoint) might be able to participate as a quasi-equal in the activities of soldiers in a span within the rank structure, as Killworth achieved in work which was focused at the platoon level (1997) and involved participating in the activities of recruits, private soldiers, junior NCOs, senior NCOs and subalterns.[10] Equally, an outsider might experience life in a single part of a unit selected by themself, as Hockey did in his study of recruits and private soldiers (1986). As an insider, of the status of commissioned officer, these were not options for me. Because of my rank (equivalent to that of a unit's commanding officer) I had no mechanism, apart from interview, by which I could capture the view from the bottom of the rank structure, so any attempt to 'study up' (Nader 1974) could only be achieved through the filter of what junior members of the unit chose to tell me. All data I gathered on important subordinate elements of the unit (the lives of private soldiers, conventions in the sergeants' mess or sub-unit bars, for example) were therefore either conditioned by my status as an officer when I gathered it directly, or, when it was indirectly gathered, emerged through interview or casual conversation with soldiers and non-commissioned officers. I had to accept that the circumstances in which indirect data were gathered were affected to some degree by the relative differences in rank. I was always 'studying down', with all the disadvantages that this brings (Womack 1995).

There was, however, one significant mitigating factor. This had to do with the precise nature of my insider status when visiting units for research. As we saw earlier, I was conducting *anthropology from within* with respect to the overall institution of the British Army, but after 1986 I was conducting *anthropology of the familiar* as I had no formal position in the units I visited. This meant that, in spite of my rank (by then, lieutenant colonel) I was not in the chain of command of any of the soldiers I interviewed. I had no formal authority to give orders to or to make life difficult for any of the interviewees. And given the strong bonding within British military units both I and my interviewees knew that, should I become awkward, the unit chain of command would be more likely to support its own soldiers than me as an outsider, irrespective of my senior rank. This factor somewhat diminished any threat of dominance that I might have brought with me.

The first, and overriding, research stratagem which I employed in interacting with soldiers was to make no denial of my rank or true status, but rather to be open about them at all times in an attempt not to compromise any basis of trust.

Second, I adopted a form of dress which did not proclaim my rank, but did not deny my commissioned status either. This was the culturally recognizable 'uniform' of the officer-out-of-uniform (tweed jacket, shirt, tie and light brown or grey trousers). I topped off this ensemble with half-eye reading glasses. This appeared to present a non-threatening image that was culturally coherent with the group I was studying. My dress was 'appropriate' without projecting organizational power or authority and it contained no badges of rank. The glasses also conveyed a studious air which did not fit readily with the cultural paradigms of rank, power and masculinity. Deep cultural immersion in the *habitus* of the Army therefore permitted me to project a relevant and effective *bodily hexis* (Bourdieu 1990: 66–79, etc.)

Third, I was careful to create situations as far as I could which put the subjects in a structurally superior position, whatever their rank. For example, when interviewing individuals I was often given the use of an empty office and I always invited the interviewee to sit behind the desk while I sat somewhere else (as often as possible in what might be called a 'client's chair' at a lower level in front of the desk) and the interviewee always had control of the tape recorder.

Another important stratagem was to make sure that all interviewees knew what they were being asked to do (talk freely about themselves and their lives), what I was going to do with the information they were giving me, and the degree to which I was going to protect that information (in other words I ensured their informed consent). In the latter respect I promised them that I would never reveal their identities in quoting what they told me and I would protect my notes and the tapes.

Finally – and I believe that this was highly significant – I expressed profound interest in what all of the participants were saying, and took care to remain resolutely polite. These were hardly an effort on my part and it seemed to make a significant impression on the speakers, most of whom seemed to warm to the task of talking about their lives.

It is impossible to tell how much trust was engendered by these actions on my part, but it is certainly the case that several interviewees expressed opinions that broke the convention of silence on sensitive topics in the face of strangers (which implies a significant degree of trust).

It is my view that these measures somehow created a special context in which differences in rank were accepted but did not dominate the interview encounters. One particular phenomenon that confirmed me in the view that the context was special was that in all but a very few cases the interviewees dropped the use of the profanities and swearwords that are a normal part of all verbal exchanges within the organizational culture of the Army (words that have been perceptively called 'sentence enhancers' by Dorn *et al.* (2001)). This was a considerable surprise, as I had made no attempt to suppress the use of these words. Non-military commentators have suggested that the absence of sentence enhancers indicated a dominant/subaltern context in the relationship between interviewer and interviewee, but this shows no understanding of Army culture: the words are as common in interchanges across ranks as they are in peer-to-peer contexts.

Once gathered, data had to be managed and analysed. It is self-evident that data management and analysis are important elements in any research, particularly in the case of social science because the data are so varied and unstructured. For the insider anthropologist, however, there are some acute issues. As we have seen, both data collection and analysis will almost certainly be affected by preconceptions arising from the researcher's own cultural baggage, in spite of all attempts to take an ersatz detached view. And it seems to me that the more the study resembles *anthropology from within* the more pronounced these effects will be.

I used two particular techniques to hedge against these difficulties. The first was to collect as much data as possible directly from informants – to allow them to speak for themselves in as unhindered a way as possible (as we saw above) – and rigorously to separate my analytical comments from these data. In all field notes and interview transcriptions I used different fonts for data and for comment, starting each comment statement with the word 'COMMENT' in block capitals. This had the advantage of making the two types of text distinct and allowing me subsequently to look for bias and discourses of power in my own analysis. It also permitted an auditable trail of evidence as to the type, source and assigned significance of all data. The second technique was to take a version of the 'grounded theory' approach advocated by Glaser and Strauss (1968). All theoretical conclusions and elements arose solely from the data: no data, no theoretical construction. Although of course there was always the risk that the data itself were culturally conditioned, this ensured that at no point did I jump to an unjustified conclusion or declare as fact something which I felt must be 'right' in the absence of data to confirm it. All this was made much easier by the use of an unstructured qualitative data handling computer package[11] that allowed data to be ordered, sorted, combined and recombined, and allowed theoretical elements to be controlled as they emerged.

Conclusion

How, then, can this chapter contribute to the wider issue of insider anthropology in general? Both my experience as researcher and the surrounding theoretical literature indicate that any degree of insider status will bring difficulties for the anthropologist in achieving the scientific detachment advocated in traditional anthropology. The researcher cannot be a stranger and an insider at the same time. They are likely to overlook aspects of the field that appear so mundane and ordinary that they do not provoke attention, and, moreover, their findings are likely to be contaminated by local cultural bias and distortion. This implies that the insider researcher needs to be self-consciously disciplined and methodical in making observations and in data recording and analysis to allow for the detection of cultural artefacts in their research. Although it would probably be an advantage in any social study, methodological procedures such as the strict separation of data from comment and analysis, and the provision of a clear trail of evidence from data to theory and conclusions, are likely to be vital in the fight against the effects of too close involvement with the target culture. This may be enhanced by such techniques as self-consciously trying to think like a third party – to create an ersatz 'stranger' point of view, as we saw in the work of Collins (2002).

On the other hand, the insider anthropologist can take complementary advantages from their pre-knowledge of the culture of the group of interest. They can make culturally sensitive connections from an early stage between the various data sets and communicate with members of the group in a way that chimes with their attitudes, assumptions and expectations. When it comes to analysis, they are in a position to identify cultural nuances that an outsider might miss.

A second issue is to do with the varied nature of 'insider anthropology'. This study has shown that the expression is not useful on its own because it lacks any precision, covering with a single term what is in fact a range of research situations. The typology used in this chapter provides a sharper conceptual tool. Different degrees of insider-ness bring with them different advantages and challenges. It is important that the researcher identifies where their research lies within the range of insider-ness and works out techniques for managing the resultant advantages and challenges. This study has also shown that the degree of insider-ness, and the advantages and challenges that go with them, may also vary in different contexts within the same overall research envelope. For example, although on the grand scale my project was *anthropology from within* at an institutional level, in the practical research context it began with *anthropology from within* but progressed to *anthropology of the familiar*, with different implications in terms of access, power relationships, cultural assumptions, and so on.

Some practical aspects for wider application also emerged from this study. In particular, the researcher can gain considerable advantage using their insider knowledge of the culture by projecting a culturally relevant *habitus* and *bodily hexis*. In my case, for example, it seems that wearing carefully selected culturally appropriate but non-threatening clothing, and the techniques employed to reduce the impact of my rank, were important elements for the success of the research. In parallel with such measures, I would advocate scrupulous honesty, careful explanation of what is taking place (to ensure informed consent) and, whenever possible, allowing interviewees to have as much control of the interview situation as possible. Such measures can do much to promote trust, a vital element in qualitative research.

Finally, let it be said on the evidence of this study that insider anthropology, although much against the original tenets of conventional anthropology, is not only possible, but can be engaging and exciting. If knowing 'the other' is important – and we social scientists believe that it is – then knowing oneself can be equally so.

Notes

1 A 'battery' is a sub-division of an artillery regiment commanded by a major and comprising approximately 100 soldiers of all ranks.
2 The sergeants and warrant officers comprise the senior non-commissioned element within a unit and provide the middle layer of command and management.
3 The 'Sergeants' Mess' in a barracks is a building (formally known as the Warrant Officers' and Sergeants' Mess) which provides exclusive accommodation, eating and leisure arrangements for the warrant officers and sergeants of a military unit or station.
4 The Battery Sergeant-Major is the senior of three Warrant Officers Class Two (WO2) in an artillery battery, and is the senior non-commissioned individual. In this case, he is the representative head of the Battery's sergeants and warrant officers.
5 All names are pseudonyms.

6 'Soldier' in this chapter refers to any member of a military unit, regardless of rank or position in the unit. Thus the term covers both the Commanding Officer and the most newly arrived private soldier.
7 At the time of the research, the 'combat arms' comprised those parts of the Army that are organized and trained to fight in direct contact with the enemy in formed units in conventional war. They included, for example, the Royal Armoured Corps, the Royal Artillery and the infantry, but not the Royal Army Medical Corps, the Royal Logistic Corps or the Royal Electrical and Mechanical Engineers. The categories have since been redrawn into a three-fold system of 'combat arms, combat support arms and combat service support'.
8 A soldier at 'regimental duty' is serving on the strength of an operational or training unit.
9 A unit is a self-contained group of between four and eight hundred soldiers commanded by a lieutenant colonel and comprising a number of sub-units commanded by majors.
10 'Subalterns' comprise junior officers (the first two commissioned ranks of second lieutenant and lieutenant, or regimentally defined equivalents).
11 The particular package was Sage's NUD*IST 4.

References

Beattie, J. (1966) *Other Cultures: Aims, Methods and Achievements in Social Anthropology*, London: Routledge & Kegan Paul Ltd.

Bourdieu, P. (1990) *The Logic of Practice* (translated by R. Nice), Cambridge: Polity Press.

Burawoy, M., Gamson, J., Burton, A., Ferguson, A., Salzinger, L., Ui, S., Hurst, L. and Gartrell, N. (1991) *Ethnography Unbound: Power and Resistance in the Modern Metropolis*, Berkeley: University of California Press.

Cerroni-Long, E. L. (1995) *Insider Anthropology*, Arlington, VA: National Association for the Practice of Anthropology.

Clifford, J. and Marcus, G. E. (1986) *Writing Culture: The Poetics and Politics of Ethnography*, Berkeley: University of California Press.

Collins, P. (1998) 'Quaker Worship: An Anthropological Perspective', *Worship*, 72, 501–515.

Collins, P. (2002) 'Connecting Anthropology and Quakerism: Transcending the Insider/Outsider Dichotomy', in E. Arweck and M. D. Stringer (eds) *Theorizing Faith: The Insider/Outsider Problem in the Study of Ritual*, Birmingham: University of Birmingham Press.

Dorn, W., Tibbitt, P. and Williams, M. (2001) 'Sailor Mouth', *SpongeBob SquarePants*, Episode 38a. USA.

Fleck, J. (1994) 'Knowing Engineers?: A Response to Forsythe', *Social Studies of Science*, 24: 105–113.

Forsythe, D. E. (1992) 'Blaming the User in Medical Informatics: The Cultural Nature of Scientific Practice', in D. Hess and L. Layne (eds), *Knowledge and Society. Volume 9. The Anthropology of Science and Technology*, Greenwich, CN: JAI.

Forsythe, D. E. (2001) *Studying Those Who Study Us: An Anthropologist in the World of Artificial Intelligence*, T. Lenoir and H. U. Gumbrecht (eds) Stanford, CA: Stanford University Press.

Fox, K. (2004) *Watching the English: The Hidden Rules of English Behaviour*, London: Hodder & Stoughton.

Glaser, B. G. and Strauss, A. L. (1968) *The Discovery of Grounded Theory: Strategies for Qualitative Research*, London: Weidenfeld and Nicolson.

Goldschmidt, W. (1995) 'The Unfamiliar in the Familiar', in E. L. Cerroni-Long (ed.) *Insider Anthropology, NAPA Bulletin 16*, Arlington, VA: National Association for the Practice of Anthropology.

Hockey, J. (1986) *Squaddies: Portrait of a Subculture*, Exeter: University of Exeter.

Irwin, A. (2002) 'The Social Organization of Soldiering: A Canadian Infantry Company in the Field', PhD, Manchester: Manchester University.

Jaffe, A. (1995) 'The Limits of Detachment: A Non-Ethnography of the Military', in E. L. Cerroni-Long (ed.) *Insider Anthropology, NAPA Bulletin 16*, Arlington, VA: National Association for the Practice of Anthropology.

Killworth, P. (1997) 'Culture and Power in the British Army: Hierarchies, Boundaries and Construction', PhD, Cambridge: Cambridge University.

King, A. (2006) 'The Word of Command: Communication and Cohesion in the Military', *Armed Forces and Society*, 32: 493–512.

Kirke, C. (2002) 'Social Structures in the Regular Combat Arms Units of the British Army: A Model', PhD, Shrivenham: Cranfield University.

Kirke, C. (2009) *Red Coat Green Machine: Continuity in Change in the British Army 1700 to 2000*, London: Continuum.

Labaree, R. V. (2002) 'The Risk of "Going Observationalist": Negotiating the Hidden Dilemmas of Being an Insider Participant Observer', *Qualitative Research*, 2: 97–122.

Liebow, E. (1995) 'Inside the Decision-Making Process: Ethnography and Environmental Risk Management', in E. L. Cerroni-Long (ed.) *Insider Anthropology, NAPA Bulletin 16*, Arlington, VA: National Association for the Practice of Anthropology.

Lieven, M. (1999) 'A Victorian Genre: Military Memoirs and the Anglo-Zulu War', *Journal of the Society for Army Historical Research*, 77: 106–121.

Lofgren, O. (1987) 'Deconstructing Swedishness: Culture and Class in Modern Sweden', in A. Jackson (ed.) *Anthropology at Home*, London: Tavistock Publications.

Nader, L. (1974) 'Up the Anthropologist: Perspectives Gained from Studying Up', in D. Hymes (ed.) *Reinventing Anthropology*, New York: Vintage Books.

Okely (1983a) 'Why Gypsies Hate Cats but Love Horses', *New Society*, 63(17): 251–253.

Okely, J. (1983b) *The Traveller-Gypsies*, Cambridge: Cambridge University Press.

Sanger, J. (1996) *The Compleat Observer? A Field Research Guide to Observation*, London: The Falmer Press.

Scheurich, J. J. (1997) *Research Method in the Postmodern*, London: The Falmer Press.

Womack, M. (1995) 'Studying Up and the Issue of Cultural Relativism', in E. L. Cerroni-Long (ed.) *Insider Anthropology, NAPA Bulletin 16*, Arlington, VA: National Association for the Practice of Anthropology.

Young, M. (1991) *An Inside Job: Policing and Police Culture in Britain*, Oxford: Clarendon Press.

3 Studying cross-cultural competence in the military

Methodological considerations of applied contract research for the German Armed Forces[1]

Phil C. Langer and Carsten Pietsch

Introduction

Empirical research that aims at getting differentiated insights into the military is usually determined by a dilemma situation: for those researchers who work outside the military (e.g. at civil universities), field access is often difficult to gain; and for those who work inside the military (e.g. at research institutes of the armed forces), scientific independence to choose research topics and methods autonomously and publish the research results freely may be restricted by institutional demands.

In Germany, social science research on the military basically follows the second path: it takes place as contract research on behalf of the government at the Bundeswehr Institute of Social Sciences (SOWI). The institute possesses a unique selling proposition in the field of empirical military sociology in the country (Klein 2005; WR 2009: 21–22, 54).[2] This fact raises some serious questions with methodological implications: To what extent do the frame conditions determine research methodologies, designs and processes? How does research vice versa impact its field? And which strategies can be implemented to guarantee ethically responsible research?

On the basis of a research study on cross-cultural competence in the military, which was carried out at the institute in 2010, these questions are being discussed in the following chapter. It aims at presenting a critical and self-reflexive analysis of the very conditions of research from inside the armed forces and their methodological implications. First, the institutional frame of the project and the project itself will be outlined (sections 2 and 3). Then the particular challenges of applied contract research for the armed forces will be analysed (section 4). Finally, strategies of dealing with these challenges as well as perspectives that disclose the potential of qualitative research in military contexts will be put up for discussion (section 5).

The chapter develops the following line of argument: many methodological challenges we faced in the course of our study are not specific to the military, but occur in other contexts and areas of social research and with regard to other research issues as well. Therefore, innovative research strategies that have been implemented in these areas (like migration or health studies) are worth being taken into account, if one is to conceptualize and conduct research in and for the armed forces. A decisive difference, however, concerns the impact of applied contract research for the military on the research field, notably the studied subjects themselves – the soldiers – because empirical findings may have direct consequences on and in what is commonly referred to as a "total institution" (Goffman 1959; see also Apelt 2004, 2008; Nesbit and Reingold 2008; Krainz and Slunecko 2011). This leads to a particular ethical responsibility of researchers who have to be aware of the regulative and normative implications of their findings, as Biehl (2010) points out. In this sense qualitative research offers a particular chance not only to reflect on its consequences in the field but to carry out this reflection systematically, together with the research subjects, the soldiers, and therefore acknowledge them as reflexive agents in the research process (Kühner and Langer 2010).

The institutional research context of the project

The SOWI is a national research institute in the realm of the Department of Defence. It is part of the German Armed Forces. The institute is politically mandated to carry out applied contract research related to the defined needs and interests of the armed forces and further basic research in the field of military sociology. Its task is understood by the Department of Defence in terms of providing empirical data, analytic expertise, and evidence-based recommendations for the government as supporting and legitimizing stakes and arguments in the political and organizational decision-making process. Against this background the state of the institute is commonly recognized as a double-bind situation between the institutional integration into the Department of Defence with its concrete demands, restrictions, political agendas and internal power games on the one hand and the discursive positioning in the scientific community that is ensured by the constitutionally founded freedom of research in Germany – which as well goes along with other specific demands, restrictions, political agendas and power-games on the other hand (Barlösius 2008; Langer 2009a; concerning the academic demands see also Božic 2009).[3] Barlösius (2008) notes:

> Generally, they are taken to be government agencies whose research follows political decisions. They are, therefore, considered to be part of the field of science as well as that of government.... According to the ministerial view, government research agencies are subordinate to the primacy of politics. Scientific standards are more or less considered to be fulfilled when the expertise reached stands the test of political debate. From the scientific point of view a basic requirement for

"good performance" of government research agencies is "excellent research"; due to this priority of scientific criteria, research done by government agencies does not represent a specific kind of research. From the government agencies' perspective, "best practice" depends on the (political) implementation of their research results; this is what they consider their specific feature.

In addition to the Department of Defence as government agency and the scientific community, a third point of reference can be added, which structures the research context of the institute: the public – and especially the media, which dominates the public/published opinion – that deeply influences and triggers the agenda setting in the ministry.[4]

The SOWI is one of currently 53 national research institutes in Germany that altogether have about 19,000 employers and an annual budget of approximately €1.7 billion, i.e. over 27 per cent of the staff and nearly 24 per cent of the overall budget of the German scientific landscape (Barlösius 2008).[5] In this respect research for the government forms a considerable part of the scientific research field in Germany. Most ministries maintain their own research institutes that aim at gaining knowledge with a direct reference to as different spheres as economy, health, ecology, agriculture, nutrition or the military. Their data and recommendations are used in the political decision-making process, e.g. by evaluating regulation tools for the labour market, investigating the agency of new medical therapies, or defining criteria for the national climate policy (see Hennecke 2005, 2006). In addition, they set up obligatory technical norms and offer socially and economically useful services, for example with regard to the weather forecast.

The national research institutes have been established in several waves since the last third of the nineteenth century. The changing field of government-based research reflects the social, economic and cultural developments that are politically perceived as decisive and lasting challenges. The foundation of the Federal Agency for Environment in the 1980s, e.g. points at the increasing significance of environmental issues like global warming for the political discourse and the national interest in the previous decade. The political decision to found the SOWI as a national research institute within the armed forces in 1974 has to be seen in the context of the establishment of universities of the armed forces: the institute had to provide scientific expertise to develop curricula for the education of military officers and was seen as a tool for further democratic control of the armed forces as part of society.[6]

The German national research institutes have some characteristics in common: their *status* as a government agency allows the respective ministry to directly organize, control and intervene in the research activities, which may imply orders to carry out (or stop) particular projects, to use (or to reject) certain methods and to publish (or to keep internal) results;

their *mission* to carry out research on behalf of the government can legitimize the confidentiality of finding in the national interest; their guaranteed *funding* by the ministries made them independent from external influences by other contracting bodies;[7] and the practical *applicability* of their research are fostered by the involvement of policy-makers and stakeholders of the ministries in the process of the setting-up of research agendas and conducting research projects. In the last respect it follows broader developments in the social sciences that – in opposition to the popular topos of the "ivory tower", which is commonly cited to discredit scientific work that is seen as self-referentially detached from social reality – claims for a social profitableness of research and instrumental usability of its results (see Carrier *et al.* 2007; Weingart and Schwechheimer 2007).

In this sense SOWI carries out e.g. studies on public opinion and societal attitudes towards the armed forces to actively govern the recruitment of personnel, on the ethical implications of a state-commanded use of force with regard to the education of soldiers as democratically responsible members of society, and a survey of the experiences and motivation patterns of German soldiers in Afghanistan to improve the efficiency and effectiveness of ongoing peace-keeping operations abroad.

The research study

The study this article refers to is part of a wider project on cross-cultural issues in the armed forces that was initiated against the background of an increasing German involvement in international military missions since the end of the 1990s (Berns and Wöhrle-Chon 2004). The importance of the subject was emphasized due to the identified necessity for deployed soldiers to effectively deal with other cultures – military ones in multinational co-operations as well as the cultures of the deployment regions that are perceived as "foreign" as the Afghan one for example; the programmatic aim of "winning the hearts and minds of the people" as precondition of a sustainable mission's success may be well known (Langer 2012). In addition, the recognition of the German society as a migration society is reflected in the project by looking to the cultural, ethnic and religious differences within the armed forces themselves (Menke *et al.* 2011). In this respect the project reflects the increasing importance of cross-cultural issues in research on the military in other countries (see e.g. Azari *et al.* 2010; Hajjar 2010; Tomforde 2010; Vuda 2010; Haddad 2011).

Within the project different dimensions of cross-cultural experiences and competence in the German military are addressed. First, in order to get a picture of the state of diversity management in the armed forces, experiences of stigmatization and discrimination due to ethnic and religious backgrounds are being examined. Second, an intervention study is part of a broader evaluation process of cross-cultural education to assess the impact of training methods. And third, as part of another SOWI study called "ISAF 2010" on the German deployment in Afghanistan the effectiveness of cross-cultural competence is being analysed. The present article deals with the challenges of doing research on cross-cultural competence within the latter study.

The research project "ISAF 2010", which has been carried out on behalf of the Department of Defence since February 2010, aims at examining the attitudes and perceptions of German soldiers in Afghanistan on a broad spectrum of topics like motivation, attitudes toward political mandate and military mission, experiences of violence, health behaviour and cross-cultural competence. The study team[8] accompanied the 22nd German ISAF contingent, which was deployed to the northern part of Afghanistan (Regional Command North) by the majority from March to November 2010, using quantitative and qualitative methods (questionnaire survey, semi-structured interviews and group discussions). In order to analyse developments that may occur over the time of deployment, three different phases for surveying the soldiers with these methods (shortly before, within and about six weeks after the deployment) were chosen. In this respect the study team took part at pre-deployment trainings and education seminars of the 22nd contingent and visited the contingent during a four-week field-research in Kunduz and Mazar-e-Sharif in the North of Afghanistan in May 2010, where 163 formal interviews and group discussions were realized; it accompanied units outside the camps, e.g. during patrol and intelligence missions and the education of Afghan security forces by the military police. Finally, after returning to Germany, post-deployment seminars were examined.

The qualitative part of the study was designed and conducted in the tradition of Grounded Theory approaches (Strauss and Corbin 1998). The interviews and group discussions followed a manual that was constructed to cover the essential dimensions of the deployment reality of the soldiers. Following the idea of the interview and group discussion as a "naturally" occurring and developing talk[9] and trying to avoid sticking too closely to the manual (and therefore losing an empirical openness in research), the manual was seen rather as a loose point of reference than a obligatory tool one has to stick to (see Hopf 1978). However, "cross-cultural competence" was defined as a decisive issue in the manual. It was considered systematically in all research encounters. Consequently, the interview manual included questions about cross-cultural significant situations soldiers experience in order to fulfil their mission, their perception of cultural "otherness", and their assessment of culture-related training before their deployment. The governing research question was: Are German soldiers adequately trained to deal with the different/alien cultural context in the Afghan theatre?[10]

The following arguments refer to a reflection of challenges that occurred during the research process and are based primarily on the qualitative research experiences.

Research challenges

The research challenges that went along with our study of cross-cultural competence in Afghanistan can analytically be distinguished with respect to three aspects: the research topic itself (cross-cultural competence), the kind of research (applied conduct research) and the institutional context of research (the armed forces).

First, the research topic of cross-cultural competence poses a methodological challenge that can be referred to as "dilemma of difference". Any research that aims at studying experiences of difference such as cultural and ethnic ones runs either the risk of discursively producing the very differences one is interested in by its own research design (Badawia *et al.* 2003) – e.g. by presupposing a certain "otherness" of the identified "other" – or fearfully avoiding othering and essentialist ascriptions – and thus underestimating both experiences of "being different" and of "being seen as different". Several of these dilemmas have so far been critically addressed in the context of a politics of recognition (Taylor and Guttman 1994; Benhabib 1996) and cross-cultural pedagogy (Kiesel 1996). Kühner and Langer (2010) have recently discussed this phenomenon of a "methodological othering" in qualitative migration research.

What does that mean for our study of cross-cultural competence in the ISAF mission? In our research question that asked about soldiers' coping with cultural "otherness" in theatre, the mentioned dilemma was present from the very beginning. It presupposed a constitutive cultural "otherness" of the deployment context, the civil population and the Afghan national security forces, that the soldiers have to work with as part of the so-called "partnering" strategy (see Alford and Cuomo 2009; Cordesman *et al.* 2010; Carter and Alderson 2011). From a methodological perspective one could argue that this, at least to some extent, necessarily implies an adoption of cultural stereotypes about the country and its people that makes it difficult to assess, when soldiers' statements about cultural differences in the interviewee can be understood as a sign of cross-cultural sensitivity or a hint at unconscious announcement of prejudices. How can this question be answered without falling into a positivistic essentialism (see Schneider 2010: 430–432)?

In particular, these challenges became obvious during the interviews with cultural "insiders": we asked Afghan translators, who worked for the German armed forces, about their assessment of the cross-cultural competence of German soldiers. In doing this, however, we performatively made them "others" and set up a clear difference between the German, "Western", Christian and the Afghan, "Oriental", Muslim culture.[11] The translator's resistance in answering our questions that became obvious in the failure of every attempt to carry out a formal, manual-led, standardized interview, was therefore significant. We were not given the opportunity of conducting formal interviews with the translators throughout the project. Instead they invited us to common dinners and involved us in informal talks about our perception of the mission and the Afghan culture more than once. In this sense, we were presented a mirror that reflected our presumptions and turned around the power relation in the research encounter. Reflecting our irritation, frustration, even annoyance about this development as a counter-transference in its psychoanalytic meaning (see Devereux 1967) we became aware of the "blind spot" in our approach: the presumption of a fundamental cultural difference that was inherent in our questions about the "other" and that ignored obvious similarities (like a common socialization into a military culture that the German soldiers shared with soldiers of the Afghan National Army) as well as further (socioeconomic, gender, age...) differences that influence perception of and interaction in social reality. We were, therefore, confronted with the "dilemma of difference" mentioned above. However, as for any dilemma, there cannot be a best research practice that may claim to solve it. The only way to cope with this dilemma seemed to be "reflexivity" – a research attitude which seems increasingly appreciated in qualitative methodology (Macbeth 2001; Breuer *et al.* 2002; Finlay 2002; Mruck and Breuer 2003; Guillemin and Gillam 2004). Adopting reflexivity with regard to our research question led to a modification in the research strategy: instead of asking, to what extent one feels prepared to act in the "other" cultural setting in theatre, we explicitly addressed the discursive production of "otherness" in pre-deployment cross-cultural trainings in the interviews.

The methodological challenge that unfolded with regard to the research topic of cross-cultural competence is not specific to the military – it rather hints at the importance to systematically integrate self-reflexive tools in the research design. As will be argued in the final section of this chapter, the qualitative paradigm offers a privileged frame for this.

Second, the kind of research which can be characterized as applied contract research, posed further challenges with methodological implications. It is *per definitionem* determined by a particular interest of the institution for which it is carried out. The study on cross-cultural competence in the 22nd German contingent was aimed at fostering soldiers' mission-performance in a certain cultural setting and therefore securing the overall mission's success. In this sense applied contract research calls for a methodological perspective that takes into account the concrete institutional dissemination and practical implementation of its results. It usually implies a narrow focus on the research subject and the available methods. In this context Barlösius (2008) offers a sceptical view about innovative methods that the scientific discourse would require to address certain issues. It would be quite

> a risk, if government-based research refers to the "top of science" because the methods and interpretations that are used there are often disputed within the scientific community. Such research results may

> be subject of a scientific quarrel that – when applied to political advisory – may lead to a controversy between experts. Hence, they would rather block than support political action. In order to reduce this risk it seems to be politically advisable from a government perspective to rely on unquestioned methods that belong to the established scientific tool-kit.[12]

In the case of the study "ISAF 2010" the "established scientific tool-kit" involved manual-led interviews and group discussions; all methodical instruments had to be authorized by the Department of Defence in advance and underwent a process of adjustment. The required choice of methods resulted in a methodological challenge because these "accepted" methods could not comprehend the research question about the effectiveness of cross-cultural trainings for the ISAF mission. Interviews and group discussions produce data that is constructed in a very specific, artificial context and that offers a retrospective view on events that reflect a subjective sense-making, social expectations and desirability as well as impression management dynamics. Yet, examining the significance of cross-cultural competence in theatre requires innovative methods to look at the effects of trainings in concrete actions and settings. Audiovisual recording techniques that capture situational actions and non-verbal action patterns might therefore be adequate. The systematic use of visual data is a development in qualitative social research that has taken place for the last ten years only (see e.g. Harper 2003; Emmison 2004; Knoblauch *et al.* 2008) and does not belong to the canon of established methods defined by the ministry.

Again, in the concept of reflexivity qualitative research contained a fruitful option to deal with this challenge. It directed our view on the effects of the use of our methods on the discursive production of data and led to the inclusion of appropriate interpretation strategies. The adoption of psychoanalytical approaches (see Schülein and Wirth 2011) in the course of the study, for example, helped to uncover the collective projections that were inherent in the interviews about the cultural "other".[13] The analysis of these projective drives in the discourse on cross-cultural competence allowed for a deeper understanding of the non-intended side effects that are generated in cross-cultural trainings and that may influence military actions in theatre in a counterproductive way.

Endeavours of organizing and controlling a narrow research frame, especially strategies to influence the research design in order to guarantee an "easy" application of results in the interest of the contracting body are characteristic for contract research in general. One could only say that they are displayed *in extremis* in the field of the armed forces, i.e. that they can be studied here in a paradigmatic way. Thus, methodological challenges dealing with the kind of research are, again, not specific to the military. Nevertheless the methodical restrictions that applied contract research brings about may affect the very possibility of dealing adequately with the methodological challenges that often occur in the course of a research project. Qualitative research though offers strategies to integrate these effects reflexively in the data analysis.

Third, the institutional frame of research – defined as the national research institute within the Department of Defence – produced significant methodological challenges. The research is part of the military dispositive. Understanding the military as a "total institution" it is obvious that research on behalf of the armed forces always implies the possibility of a direct impact of the research results on the field. The results can be implemented in the education and training of soldiers, therefore influencing their professional identity and self-perception.

In this respect, research in and for the armed forces can be distinguished from other fields of government-based research in Germany as an exemplary comparison with the German Youth Institute, a research agency of the Ministry for Family and Youth Affairs shows. The institute, too, affects decisions of the ministry by its research, e.g. concerning supporting measures for vulnerable young people, but the effect of the research-based decisions on the field ("the youth") are obviously quite limited and mediated through many different actors and institutions, e.g. the familial socialization, the media or non-governmental organizations that work with vulnerable adults. The social field covered by the institute is much more open and determined by multiple factors and agents than the military/armed forces and therefore the question of which interventions may be promising to best implement the political decisions are part of the research itself.

The integration of research results in military trainings and orders, in contrast, usually has far-reaching consequences for soldiers. The project on cross-cultural competence in the German Armed Forces has already contributed to the implementation of a military-wide binding education concept, the evidence-based set- and scale-up of trainings for peer facilitators, and the development of instruments for measuring cross-cultural competence. As a result cross-cultural competence is being promoted as a military key qualification for soldiers in theatre, a qualification that soldiers have to acquire not only for certain deployments, but for the sake of their future career in the armed forces. Hence the question of how to implement results is not a matter of principle, but – if the political decision about the use of the results is made – just a technical question.

A theoretical frame for understanding the effects of military-related research conducted by the SOWI is given in Michel Foucault's concept of governmentality. The semantic connection of "govern" and "mentality" aims at showing that examining technologies of power calls for an analysis of the underlying political rationality. In Foucault's concept, power is inevitably linked to the production of knowledge, while the execution of power under (post-) modern conditions is constitutively based on scientific knowledge. Lemke (2000a) writes:

> Firstly, the term signifies a specific kind of *representation*, i.e. government defines a discursive field in which the execution of power is rationalized. This e.g. happens through the production of terms and concepts, the specification of matters and limits, and the presentation of arguments. In this respect, government is not identical with concrete contents or particular methods, but implies a kind of "problematization", i.e. it defines a political-epistemological space, in which historical problems (may) occur and offers – possibly conflicting or opposing – strategies for their coping at the same time. Consequently the Foucaultian sense of "government" structures specific forms of intervention. A political rationality is not a pure and neutral knowledge that just re-presents a reality that is to be governed, but always already presents an intellectual modification of reality, in which political technologies can intervene. This implies e.g. apparatuses, mechanisms, institutions, and forms of justice that allow the government of its objects and subjects according to a political rationality.[14]

In this sense, government-based research related to the military (as governmental organization) can be understood as a manifestation of a will to constantly rationalize political decisions, make political leadership processes more effective and extend institutional attempts for order and control. Power, therefore, operates not as a (pure) hierarchical, repressive top-down power of the state, but includes a wide range of control techniques that applies e.g. to the biopolitical control of populations, social control in disciplinary institutions (schools, hospitals, psychiatric institutions, etc.) that get internalized by individuals (subjectivation), and one's control of the self. The strong relationship between this kind of power and processes of subjectivation as implied by the concept of governmentality (see Sellin 1984; Senellart 1995) has to be taken into account, if one is to understand the unique position of the SOWI in the German research landscape.

This is also important with regard to the project on cross-cultural competence. The project decisively shapes the training on cross-cultural competence in terms of a definition of skills and knowledge a "useful soldier" should have internalized, and is therefore a constitutive basis of subjectivating practices. The project constitutes a power-knowledge in the Foucaultian sense that brings up the possibilities of exploiting results beyond the concrete research context and poses questions about the responsibility of research.

To take up the former example again, the rejection of examining gender and sexual differences from an interdisciplinary and intersectionality perspective has certain consequences for the soldiers because it solely focuses on individual competences of the soldiers, handicaps the discussion of cross-cultural competence on a more complex structural level and disregards the interconnection of multiple social differences. Demands for cross-cultural competence in the armed forces then reflect attributions of personal responsibility of a military "entrepreneurial self" in the context of current governmentality drives. Instead of critically describing and analysing the military dispositive within its social context, contract research for the armed forces rather tends to replicate existing fields of knowledge, institutional power relations and thus supports a naïve reduplication of reality. Hence the following questions become relevant: Does research in cross-cultural competence contribute to an individual overstrain of the soldier? Are certain images of "otherness" being inscribed into soldier's identity that – from a normative perspective – have to be deconstructed? Are the armed forces as an organization not being freed of the necessity to reflect cultural issues in the political and tactical decision-making process by focusing on the cultural competence of the individual soldier as concrete action ability (see Tomforde 2009)?

The point here is that the institutional frame of research does not pose methodological challenges per se, but sharpens and deepens those challenges that were illustrated with regard to the topic and the kind of research. It adds a decisive ethical dimension, simultaneously underlining that ethical considerations are constitutive for any methodological reflection in doing research within the military dispositive.

The direct and privileged access to the field of study, the resources provided for the research, and the possible consequences for different actors in the field lead to a particular responsibility of the research at the SOWI. This calls for a systematic and continuous analysis of the research's role in the dispositive as part of the research design by integrating self-reflexive elements: Can or should a concept like cross-cultural competence be exploited for an improvement of effective and "successful" coping and possible fighting against and killing of the "other" through better understanding of her/his cultural background – while this very concept is historically and epistemologically rooted in a critique of social power relations, ethnocentric discourses and hierarchy structures to empower the "other"? How can we position ourselves as socially responsible researchers in this over-determined field? What ethical consequences do the political contextualization and the concrete impact of our research have for our work (and ourselves)?

The qualitative paradigm again offers a useful tool for answering these questions. Its ability to take into account the significance of power by means of an accompanying dispositive analysis in a post-Foucaultian tradition (see e.g. Bührmann 2004; Klemm and Glasze 2004) demonstrates the potential of qualitative research in doing research on military-related issues.

Strategies and perspectives

In the previous section we tried to argue that most challenges that our study on cross-cultural competence in theatre posed and that referred to the research topic and the kind of research are not specific to the military.

They also occur in other fields of sociological study where issues of difference are examined and methodical restrictions due to application-related contracts exist. Dealing with their methodological implications, therefore, does not require military-specific research strategies.

It may well be worth checking whether strategies that have been developed, implemented and tested in other contexts can be transferred to the study of the military. Kühner and Langer (2010) discussed some of these strategies in the context of qualitative studies on Holocaust Education and HIV prevention (see also Langer *et al.* 2008; Langer 2009b). With regard to the adoption of self-reflexive elements in qualitative studies, that we referred to in the previous section as well, they pointed out that

> the researchers are not the only ones who are able to reflect. Indeed it is imperative to take the researched subject of desire seriously as a reflexive agent that is (more or less consciously) aware of possible social and political consequences or instrumentalizations of research. To put this into practice the interviewee has to be addressed much more explicitly as someone who does not only give information (or even authentic self-revelations), but as someone who has his or her own thoughts and interpretations about the research process.
>
> (Kühner and Langer 2010: 76)

In particular they proposed, for example, to devote one passage of the research encounter to talk about the project on a meta-level with the interviewees to acknowledge their expectations and fears that are related to the research project.

Therefore, we would like to argue that addressing methodological problems in the study of the military can never be dissociated from the ethics of our research and our own responsibilities in research – in particular with regard to the implications for the subjects that are more direct and far-reaching than in most other fields of research (see also Bannister 1996). Especially in such a "powerful" institutional context like the military where research results may affect participants' lives quite directly and extensively (and sometimes even literally), it should be self-evident to systematically integrate participatory elements in our studies. The backtalk-focus group discussion that Frisina (2006) outlined in her research report on Muslim youths in Italy and that make the stakeholders an integral part of the knowledge-generation process by re-discussing the interpretative categories of the researcher together with the researched may be a promising tool for that.

The previous discussion of the challenges we faced in our research on cross-cultural competence in Afghanistan pointed at a particular strength of qualitative research. Hence, one should not only ask about the methodological implications of qualitative research in the military, but explore the fruitful use of qualitative research designs as a privileged strategy of dealing with methodological challenges that may occur in studying the military.

However, qualitative research itself is not without challenges. Especially in institutional contexts that aim at achieving "objectivity" by relying in large part on standardized quantitative approaches, the qualitative paradigm with its different criteria for "good research" like inter-subjectivity and researcher-related interpretations are confronted with scepticism. This brings about more difficulties in legitimizing decisions than referring to "hard facts" of quantitative research. As a consequence, in German military sociology qualitative methods are often seen just as accompanying preliminary tools in research projects that offer context information for the interpretation of statistical results or to develop a quantitative main study. In this sense, a lasting challenge for studying the military is to create an institutional sensitivity for the potential of qualitative research. Again, other fields of research may function as points of reference.

In health studies, for example, qualitative research has gained great importance in the last decades. With regard to the contribution of qualitative research for the health sector Green and Thorogood (2005) outline three aspects that can be transferred to the military context as well: first, they describe their use "in their ability to answer important questions that cannot be answered from a quantitative perspective" (p. 22). In this respect qualitative research can close research deficits on specific new, previously understudied topics and questions and in settings that are difficult to reach. Second, due to its interpretative drive, qualitative research does not only produce "other" data, the authors argue, but "provide[s] 'better' answers to questions located in less positivist epistemologies" (p. 22). In complex and sensitive contexts they cover more adequately the social reality and the intersubjective process of meaning production of the subjects that can be represented as partners in the research process (Eide and Kahn 2008). Finally, for policy-makers and stakeholders that practically work in the field, qualitative research is "often useful to 'sensitizing' them to patient's views" (p. 24), because they allow insights into the perception and interpretation of their clients and a comparison to one's own experiences; on a political level, qualitative studies have "the potential to provide evidence for population needs, the development of appropriate policy, and evidence for how to implement policy with health care staff" (p. 24).

With regard to the military, the impact of empirical research-based political decisions on the field calls for a sensitive research that takes into account the views of its subject. It will, however, be a long way until this perspective will be common sense in sociological research on the military in Germany.

Notes

1 The study this article refers to was carried out at the Bundeswehr Institute of Social Sciences on behalf of the German Department of Defence. However, neither the Bundeswehr Institute of Social Sciences nor the Department of Defence have had any influence on the conceptualization, writing and publication of the article.

2 A coherent sociological research on the military at civil universities, academies or further research institutes in Germany does not yet exist. This phenomenon can be understood as an effect of the ideological instrumentalization of research during the National Socialist era that led to a critical distance to the military of these institutions and a distinct uneasiness to address military issues that are feared to function as possible drivers of a normalization of military discourses. Moreover, the SOWI influences the academic sociology by participating in the MA course "Military Studies" at Potsdam University – the only course that specifically deals with the military in Germany from a sociological perspective. Consequently, there is a lack of institutional concurrence and discursive quarrel for "truth" in the field of military sociology.

3 In psychology the term "double-bind" refers to a communication pattern in which two conflicting messages are sent simultaneously. In the context of this article it is used to describe discursive conflicts that have certain methodological implications: the different systems of reference (the ministry and the scientific community) result in the necessity to legitimate the research topics, the used méthods, and the validation of the results quite differently. This is obvious with regard to the negotiation between the justification of institutional "worth" and use of a research question for the ministry on the one hand and the demonstration of its scientific significance in the disciplinary discourse on the other that often seem to be hardly compatible. The double-bind situation is even more complex due to the fact that both – the ministry and the scientific community – represent no monolithic actors, but contain different, sometimes inconsistent interests.

4 From a system theoretical perspective the media signifies an institutionally uncontrollable factor that contributes to the decision not to publish certain results. Attempts to control research findings by the Department of Defence, however, often provoke the non-intended opposite: the clash between a military habitus of confidence and the medial mission of investigation fuel a characteristic dynamic of non-publication of "sensitive" results, recherché and reporting by the media that involves assumptions about and oversimplification of these results, which leads to further fearful reservation and rejection by the military.

5 With regard to research-specific activities (i.e. without administration issues) these figures are somewhat different: research staff in national institutes accounts for about 15 per cent of the overall research staff in Germany and the budget in these institutes for 10 per cent of the overall research budget. This can be interpreted as an effect of the amount of co-ordination work with other government agencies as well as different institutional criteria for qualitatively "good" work that do not necessarily focus on efficiency.

6 There are similar national research institutions in many other countries, e.g. the *Centre des etudes en sciences socials de la défense* in Paris or the *Woskowy Instytut Badan Socjologicznych* in Warsaw (see Klein 2005). However, they have somewhat different thematic focuses that refer to the respective national political discourses. Also, most countries have different institutional foundations and therefore legal conditions for doing research: in the United Kingdom research for the government is much more privatized than in Germany, which means conduct research for the armed forces is carried out in civil (private and university) research institutes. In Switzerland, government-related research is integrated in civil universities; social research for the military, for example, takes place at Zurich University. Independent research at think tanks – like the RAND corporation in the United States – that are (co-)financed by the government has to be seen in this context as well.

7 The option to acquire external financial support for projects was legally enabled only at the end of 2010.
8 The study team consisted of Anja Seiffert (leader of the research area "Military Deployments of the German Armed Forces" and project leader "ISAF 2010"), Phil C. Langer (project leader "Cross-cultural Competence"), Carsten Pietsch (project leader "Naval Officers' Education") and Bastian Krause (military advisor to the team).
9 This idea is of course somewhat unrealistic because an interview for instance always implies "the creation of an unnatural social situation, introduced by a researcher, for the purpose of polite interrogation" (Kellehear 1996: 98). It therefore includes power relations and asymmetries between the interviewer and the interviewee.
10 The German term "fremd", which was used to describe the "different" culture in Afghanistan, contains notions of "alien" and "strange" as well.
11 For a discussion of the issue of cultural "insiders" and the challenge of positionality see Ganga and Scott 2006.
12 Translation from the German original by Phil C. Langer.
13 The narrow focus of the research design does not only pertain to the choice of "established" methods, but concerns the interplay of the research subject with other discourses as well. In the context of the mentioned problem of a "methodological othering" current research demonstrates the importance of addressing other differences within an intersectionality approach (see e.g. Crenshaw 1998; Bryant and Hoon 2006; Degele and Winker 2007; Vinz and Dören 2007). In this sense, doing research on difference characteristics like "culture" requires addressing its relationship to other socially significant differences – such as gender, class, age, sexual identity, which are strongly interconnected with cultural and ethnic differences. Integrating these sensitive dimensions – that still go along with social taboos in military contexts – into the research of cross-cultural competence in the military would exacerbate an easy application of research results. It was therefore not possible to formally (for example in the manual or within the quantitative part of the study in the questionnaire) widen the research design to intersectionality issues.
14 Translation from the German original by Phil C. Langer; emphasis in the original. Regarding the concept of governmentality see also Lemke 2000b.

References

Alford, J. and Cuomo, S. (2009) "Operational Design for ISAF", in "Afghanistan: A Primer", *Joint Forces Quarterly*, 53(2): 92–98.

Apelt, M. (2004) "Militärische Sozialisation", in S.B. Gareis and P. Klein (eds) *Handbuch Militär und Sozialwissenschaft*, Wiesbaden: Vs Verlag für Sozialwissenschaften, 26–50.

Apelt, M. (2008) "Sozialisation in 'totalen Institutionen'", in K. Hurrelmann, M. Grundmann and S. Walper (eds) *Handbuch Sozialisationsforschung*, 7th edn, Weinheim & Basel: Beltz, 372–384.

Azari, J., Dandeker, C. and Greenberg, N. (2010) "Cultural Stress: How Interactions With and Among Foreign Populations Affect Military Personnel", *Armed Forces & Society*, 36(4): 585–603.

Badawia, T., Hamburger, F. and Hummrich, M. (2003) *Wider die Ethnisierung einer Generation. Beiträge zur qualitativen Migrationsforschung*, Frankfurt am Main: IKO-Verlag für Interkulturelle Kommunikation.

Bannister, R. (1996) "Beyond the Ethics Committee: Representing the Other in Qualitative Research", *Research Studies in Music Education*, 6: 50–58.

Barlösius, E. (2008) *Zwischen Wissenschaft und Staat? Die Verortung der Ressortforschung*. Online. Available at: http://bibliothek.wzberlin.de/pdf/2008/p08–101.pdf (accessed 5 October 2009).

Benhabib, S. (1996) *Democracy and Difference*, Princeton: Princeton University Press.

Berns, A. and Wöhrle-Chon, R. (2004) "Interkulturelle Kompetenz im Einsatz", in S.B. Gareis and P. Klein (eds) *Handbuch Militär und Sozialwissenschaft*, Wiesbaden: Vs Verlag für Sozialwissenschaften, 322–331.

Biehl, H. (2010) "Kohäsion als Forschungsgegenstand, militärische Praxis und Organisationsideologie", in M. Apelt (ed.) *Forschungsthema Militär*, Wiesbaden: VS Verlag für Sozialwissenschaften, 139–162.

Božic, S. (2009) "Academic Narcissism and the Problem of Knowledge Accumulation in the Social Sciences", paper presented at the conference *The Future of Social Sciences and Humanities*, Brussels, October 2009.

Breuer, F., Mruck, K. and Roth, W.-M. (2002) "Subjectivity and Reflexivity: An Introduction", *Forum: Qualitative Social Research*, 3(3). Online. Available at: http://nbn-resolving.de/urn:nbn:de:0114-fqs020393 (accessed 2 January 2010).

Bryant, L. and Hoon, E. (2006) "How Can the Intersections Between Gender, Class, and Sexuality Be Translated to an Empirical Agenda?" *International Journal of Qualitative Methods*, 5(1). Online. Available at: www.ualberta.ca/~iiqm/backissues/5_1/pdf/bryant.pdf (accessed 30 November 2009).

Bührmann, A. (2004) "Das Auftauchen des unternehmerischen Selbst und seine gegenwärtige Hegemonialität. Einige grundlegende Anmerkungen zur Analyse des (Trans-) Formierungsgeschehens moderner Subjektivierungsweisen", *Forum: Qualitative Social Research*, 6(1). Online. Available at: www.qualitative-research.net/fqs-texte/1–05/05–1–16-d-htm (accessed 16 April 2006).

Carrier, M., Krohn, W. and Weingart, P. (2007) "Historische Entwicklungen der Wissensordnung und ihre gegenwärtigen Probleme", in P. Weingart, M. Carrier and W. Krohn (eds) *Nachrichten aus der Wissensgesellschaft: Analysen zur Veränderung der Wissenschaft*, Weilerswist: Velbrück Wissenschaft, 11–33.

Carter, N. and Alderson, A. (2011) "Partnering With Local Forces", *The RUSI Journal*, 56(3): 34–40.

Carter, S.M. and Little, M. (2007) "Justifying Knowledge, Justifying Method, Taking Action: Epistemologies, Methodologies, and Methods in Qualitative Research", *Qualitative Health Research*, 17(10): 1316–1328.

Cordesman, A.H., Mausner, A. and Lemieux, J. (2010) *Afghan National Security Focus: What It Will Take to Implement the ISAF Strategy*, Washington: Center for Strategic and International Studies. Online. Available at: http://csis.org/files/publication/101115_Cordesman_AfghanNationalSecurity Forces_Web.pdf (accessed 12 August 2011).

Crenshaw, K. (1998) "Demarginalizing the Intersection of Race and Sex. A Black Feminist Critique of Antidiscrimination Doctrine, Feminist Theory, and Antiracist Politics", in A. Phillips (ed.) *Feminism and Politics*, Oxford: Oxford University Press, 314–343.

Degele, N. and Winker, D. (2007) *Intersektionalität als Mehrebenenanalyse*. Online. Available at: www.rosalux.de/fileadmin/ls_sh/dokumente/Intersektionalitaet_Mehrebenen.pdf (accessed 2 January 2011).

Devereux, G. (1967) *From Anxiety to Method in the Behavioral Sciences*, The Hague & Paris: Mouton and Cie.

Eide, P. and Kahn, D. (2008) "Ethical Issues in the Qualitative Researcher-Participant Relationship", *Nursing Ethics*, 15(2): 199–207.

Emmison, M. (2004) "The Conceptualization and Analysis of Visual Data", in D. Silverman (ed.) *Qualitative Research. Theory, Method and Practice*, 2nd edn, London: Sage, 246–265.

Finlay, L. (2002) " 'Outing' the Researcher: The Provenance, Process, and Practice of Reflexivity", *Qualitative Health Research*, 12(4): 531–545.

Flick, U. (2006) *An Introduction to Qualitative Research*, 6th edn, London: Sage.

Frisina, A. (2006) "Back-talk Focus Groups as a Follow-Up Tool in Qualitative Migration Research: The Missing Link?", *Forum: Qualitative Social Research*, 7(3). Online. Available at: http://nbn-resolving.de/urn:nbn:de:0114-fqs060352 (accessed 12 May 2009).

Ganga, D. and Scott, S. (2006) "Cultural 'Insiders' and the Issue of Positionality in Qualitative Migration Research: Moving 'Across' and Moving 'Along' Researcher-Participant Divides", *Forum: Qualitative Social Research*, 7(3). Online. Available at: www.qualitative-research.net/index.php/fqs/article/view/134 (accessed 9 December 2009).

Goffman, E. (1959) *Asylums: Essays on the Social Situation of Mental Patients and Other Inmates*, Chicago: Aldine Publishing Company.

Green, J. and Thorogood, N. (2005) *Qualitative Methods for Health Research*, Thousand Oaks: Sage.

Guillemin, M. and Gillam, L. (2004) "Ethics, Reflexivity, and 'Ethically Important Moments' in Research", *Qualitative Inquiry*, 10(2): 261–280.

Haddad, S. (2011) "Teaching Diversity and Multicultural Competence to French Peacekeepers", *International Peacekeeping*, 17(4): 566–577.

Hajjar, R.M. (2010) "A New Angle on the U.S. Military's Emphasis on Developing Cross-Cultural Competence: Connecting In-Ranks' Cultural Diversity to Cross-Cultural Competence", *Armed Forces & Society*, 36(2): 247–263.

Harper, I. (2003) "Reimagining Visual Methods: Galileo to Neuromancer", in N.K. Denzin and Y.S. Lincoln (eds) *Collecting and Interpreting Qualitative Materials*, Thousand Oaks: Sage, 176–198.

Hennecke, M. (2005) *Über das Selbstverständnis der Ressortforschung. Presentation at the Wissenschaftszentrum*. Online. Available at: http://ressortforschung.de/de/res_medien/vortrag_he_ress_fo_beim_stifterverband.pdf (accessed 30 November 2009).

Hennecke, M. (2006) *The German Federal Research Institutes and their Role in Policy Advice*. Online. Available at: http://ressortforschung.de/de/res_medien/vortrag_he_lecture_bbaw.pdf (accessed 28 January 2010).

Hopf, C. (1978) "Die Pseudo-Exploration – Überlegungen zur Technik qualitativer Interviews in der Sozialforschung", *Zeitschrift für Soziologie*, 7(2): 97–115.

Kellehear, A. (1996) "Unobtrusive Methods in Delicate Situations", in J. Daly (ed.) *Ethical Intersections: Health Research, Methods and Researcher Responsibility*, Sydney, Australia: Allen & Unwin, 97–105.

Kiesel, D. (1996) *Das Dilemma der Differenz. Zur Kritik des Kulturalismus in der interkulturellen Pädagogik*, Frankfurt am Main: Cooperative-Verlag.

Klein, P. (2005) "Militärsoziologische Forschungseinrichtungen, Arbeitskreise und wissenschaftliche Vereinigungen", in N. Leonhard and I.-J. Werkner (eds)

Militärsoziologie – Eine Einführung, Wiesbaden: VS Verlag für Sozialwissenschaften, 346–351.

Klemm, J. and Glasze, G. (2004) "Methodische Probleme Foucault-inspirierter Diskursanalysen in den Sozialwissenschaften. Tagungsbericht: 'Praxis-Workshop Diskursanalyse'", *Forum: Qualitative Social Research*, 6(2). Online. Available at: http://nbn-resolving.de/urn:nbn:de:0114-fqs0502246 (accessed 28 April 2006).

Knoblauch, H., Baer, A., Laurier, E., Petschke, S. and Schnettler, B. (2008) "Visual Analysis, New Developments in the Interpretative Analysis of Video and Photography", *Forum: Qualitative Social Research*, 9(3). Online. Available at: http://nbn-resolving.de/urn:nbn:de:0114-fqs0803148 (accessed 14 March 2010).

Krainz, U. and Slunecko, T. (2011) "Negotiating Cultural Differences in a Total Institution: Muslim Conscripts in the Austrian Armed Forces", in I. Menke and P.C. Langer (eds) *Muslim Service Members in Non-Muslim Countries. Experiences of Difference in the Armed Forces in Austria, Germany, and The Netherlands*, Strausberg: SOWI (=*Forum International*, 29), 105–134.

Kühner, A. and Langer, P.C. (2010) "Dealing With Dilemmas of Difference. Ethical and Psychological Considerations of 'Othering' and 'Peer Dialogues' in the Research Encounter", *Migration Letters*, 7(1): 69–78.

Langer, P.C. (2008) "Aktuelle Herausforderungen der schulischen Thematisierung von Nationalsozialismus und Holocaust", *Einsichten und Perspektive*, Themenheft 1: 10–27.

Langer, P.C. (2009a) "Doing Research in the Name of War? Experiences from a Social Science Institute Within the Army", paper presented at the conference *The Future of Social Sciences and Humanities*, Brussels, October 2009.

Langer, P.C. (2009b) *Beschädigte Identität. Dynamiken des sexuellen Risikoverhal- tens schwuler und bisexueller Männer*, Wiesbaden: VS Verlag für Sozialwissenschaften.

Langer, P.C. (2012) "Erfahrungen von 'Fremdheit' als Ressource verstehen", in A. Seiffert, P.C. Langer and C. Pietsch (eds) *Der Einsatz der Bundeswehr in Afghanistan*, Wiesbaden: VS Verlag für Sozialwissenschaften, 121–140.

Langer, P.C., Cisneros, D. and Kühner, A. (2008) "Aktuelle Herausforderungen der schulischen Vermittlung von Nationalsozialismus und Holocaust. Zu Hintergrund, Methodik und Durchführung der Studie", in *Einsichten und Perspektiven*, Special Issue 1, 10–27. Online. Available at: http://192.68.214.70/blz/eup/01_08_themenheft/2.asp (accessed 30 March 2012).

Lemke, T. (2000a) *Neoliberalismus, Staat und Selbsttechnologien. Ein kritischer Überblick über die governmentality studies*. Online. Available at: www.thomaslemkeweb.de/engl.%20texte/Neoliberalismus%20ii.pdf (accessed 5 February 2010).

Lemke, T. (2000b) *Foucault, Governmentality, and Critique*. Online. Available at: www.andosciasociology.net/resources/Foucault$2C+Governmentality$2C+and+Critique+IV-2.pdf (accessed 5 February 2010).

Macbeth, D. (2001) "On 'Reflexivity' in Qualitative Research: Two Readings, and a Third", *Qualitative Inquiry*, 7(1): 35–68.

Menke, I., Langer, P.C. and Tomforde, M. (2011) "Challenges and Chances of Integrating Muslim Soldiers in the Bundeswehr. Strategies of Diversity Management in the German Armed Forces", in I. Menke and P.C. Langer (eds) *Muslim Service Members in Non-Muslim Countries. Experiences of Difference in the Armed Forces in Austria, Germany, and The Netherlands*, Strausberg: SOWI (=*Forum International*, 29), 13–42.

Mruck, K. and Breuer, F. (2003) "Subjectivity and Reflexivity", *Forum: Qualitative Social Research*, 4(2). Online. Available at: http://nbn- resolving.de/urn:nbn:de:0114-fqs0302233 (accessed 11 January 2011).

Nesbit, R. and Reingold, D.A. (2008) *Soldiers to Citizens: The Link between Military Service and Volunteering*. Online. Available at: www.rgkcenter.org/sites/default/files/file/2008papers/Nesbitt.pdf (accessed 1 October 2011).

Schneider, W.L. (2010) "Kultur als soziales Gedächtnis", in G. Albert and S. Sigmund (eds) *Soziologische Theorie kontrovers*, Wiesbaden: VS Verlag für Sozialwissenschaften, 427–440.

Schülein, J.A. and Wirth, H.-J. (eds) (2011) *Analytische Sozialpsychologie. Klassische und neuere Perspektiven*, Gießen: Psychosozial-Verlag.

Sellin, V. (1984) "Regierung, Regime, Obrigkeit", in O. Brunner, W. Conze and R. Koselleck (eds) *Geschichtliche Grundbegriffe, Band 5*, Stuttgart: Reclam, 361–421.

Senellart, M. (1995) *Les arts de gouverner. Du regimen médiéval au concept de gouvernement*, Paris: Editions du Seuil.

Strauss, A. and Corbin, J. (1998) *Basics of Qualitative Research. Techniques and Procedures for Developing Grounded Theory*, Thousand Oaks: Sage.

Taylor, C. and Gutmann, A. (eds) (1994) *Multiculturalism and "The Politics of Recognition"*, Princeton: Princeton University Press.

Tomforde, M. (2009) "Bereit für drei Tassen Tee? Die Rolle von Kultur für Auslandseinsätze der Bundeswehr", in S. Jaberg, H. Biehl, G. Mohrmann and M. Tomforde (eds) *Auslandseinsätze der Bundeswehr. Sozialwissenschaftliche Analysen, Diagnosen und Perspektiven*, Berlin: Duncker & Humblot, 124–148.

Tomforde, M. (2010) "The Distinctive Role of Culture", *Peacekeeping International*, 17(4): 450–456.

Vinz, D. and Dören, M. (2007) "Diversity Policies and Practices: A New Perspective for Health Care", *Journal of Public Health*, 15(5): 369–376.

Vuga, J. (2010) "Cultural Differences in Multinational Peace Operations: A Slovenian Perspective", *International Peacekeeping*, 17(4): 554–565.

Weingart, P. and Schwechheimer, H. (2007) "Institutionelle Verschiebungen der Wissensproduktion – Zum Wandel der Struktur wissenschaftlicher Disziplinen", in P. Weingart, M. Carrier and W. Krohn (eds) *Nachrichten aus der Wissensgesellschaft: Analysen zur Veränderung der Wissenschaft*, Weilerswist: Velbrück Wissenschaft, 41–71.

Wissenschaftsrat (WR) (2009) *Stellungnahme zum Sozialwissenschaftlichen Institut der Bundeswehr (SWInstBw), Strausberg*, Bonn: Wissenschaftsrat.

4 Evolving experiences

Auto-ethnography and military sociology – a South African immersion

Ian Liebenberg

Introduction

As far as traditional or orthodox research approaches go, this contribution differs. Despite the fact that auto-ethnography originated back in the 1990s and some of its pioneers stemmed from the USA and the UK the approach is not always well known. For some more orthodox researchers auto-ethnography strikes a jarring note and on occasion was described as "research against the grain" or "too close up and personal". Disliked by some, advanced by others, yet contemporary qualitative research can benefit from auto-ethnography and prominent authors advocate its value (De Marrais 1998; Ellis and Bochner 2000; Czarniawska 2004; Gingras 2007).

Auto-ethnography as an approach frequently deploys the personal narrative triggered by individual experience. Readers may recall an occasion or two where a reviewer critically commented that one does not see, or "feel" enough of the author and that some of the shared views were "distant" (i.e. more of an intellectual treatment of the author's experience or findings rather than the context or emotions lived through). Auto-ethnography attempts to fill this lacuna and the underlying idea is to share slices of life or immersion into the human experience. Humans are bodily or somatic beings (Hanna 1970; Luijpen 1980). Sharing experiences also in the search for knowledge thus relates to humanness, shared and personal trials, emotions and feelings. Auto-ethnography as an approach bring the "researcher back in" without losing the context or those who are studied.

Auto-ethnography needs – if not blooms on – the reflected-upon-experience of the researcher as one of the tools of research. Personal examples and reflections are crucial and add value to the narrative of the self and others within the social context.

Take for example the use of the peer debriefer, a fellow research participant acting as a consistent soundboard. Peer debriefers (those chosen by the author him/herself to critically comment on their work and share their research path) can assist the researcher greatly in the process of obtaining a doctoral degree or other qualitative research endeavours. Peer debriefers are intrinsic human tools in the auto-ethnographic process. It is imperative for the auto-ethnographer to take heed of those that accompany him or her on the research path. If we are the tools of qualitative research, especially in auto-ethnography, peer debriefers are those that co-hone and sharpen the research tools including the researcher as a tool himself. The peer debriefer's comments facilitate **re**-thinking about the **re**-search and personal experience and **re**-writing. The debriefing comments lead to **re**-flection and form part of the process of gaining and sharing insights.

It is not without reason that it is argued that "auto-ethnography [is a] particular kind of writing that seeks to unite ethnographic (looking outward at a world beyond one's own) and autobiographical (gazing inwards for a story of one's self) intentions" (Schwandt 2001: 13). "Auto-ethnography is intended to evoke, rather than state a claim. It is meant to invite the reader into the text to relive the experience, rather than to analyse it" (ibid.). It is an invitation to dialogue on humaneness rather than the cold statement of hypotheses and truths cast in iron. For orthodox researchers or those with positivist inclinations auto-ethnography may perhaps strike a jarring note.

The challenges of auto-ethnography

Burnier (2006: 412) suggests that,

> [personal] writing is hybrid in character and combines an individual's story with his or her [the author's] personal story. It is writing that is not strictly scholarly because it contains the personal, and yet it is not strictly personal because it contains the scholarly ... it erases the false dichotomy between scholarly and personally.

I have to share a note of caution here with the reader. Auto-ethnography is no less rigorous as a research approach compared to other research approaches. The researcher has to deal with facts, dates and utterances of research participants with as much caution as any other researcher. Research ethics and the norms of the social scientific community remain as important as in any other project. Terminology such as triangulation and verifiability are not used. However, the chosen research process should still be replicable, transferable and trustworthy. The researcher or human tool on the research path (in other words, I as author) should be able to provide a research audit trail and be honest about the steps taken and the shortcomings of the study (the latter which after all, is but a critical extension of the researcher as person). The completed process should provide pointers on both the research process and its social value for others in other settings.

The involvement of the individual in any social process is not that of "being apart" from the process but "being part". Social and individual experiences inform each other. In reporting on a research process and findings, personal involvement is present (Plummer 2001: 34–35). Life experiences are not devoid of possible insights that are valuable to a growing corpus of knowledge. No person is an island and collective experiences can bring about thoughtful insights leading to more accommodative future human relationships – even inform policy choices. Auto-ethnography as a knowledge-generating exercise is relevant in this context.

We cannot divorce ourselves from our research or its findings in the search for knowledge. Nor can I do so in sharing this research narrative with you as the reader. C. Wright Mills (1972: 7ff.) argued poignantly that researchers cannot distance themselves from the findings or the research process. Qualitative elements have an important place. Essentially the researcher is *somatic* or *bodily* being (Hanna 1970). He is an onto-ethical body (Peperzak 1977). I am, as researcher, an individual growing in knowledge, learning from mistakes, gaining insights and enmeshed in interactive social processes (see Higate and Camron 2006: 220; also compare the existentialist and phenomenologist Luijpen (1980) on the involvement of the subject).

Since the 1990s, authors within the new genre of auto-ethnography have argued convincingly for the acceptance of "qualitative subjectivity" in broadening our basis of social knowledge (Ellis and Bochner 1996, 2000; Garrat 2003; Denzin 2006: 419ff.; Etherington 2006). The acknowledgement of the above provides one of the main challenges for those who embark on auto-ethnography.

Choices, choices and choices: taking up a doctoral thesis

I embarked on a doctoral thesis some years ago. In between other projects, political activism, the love for writing and lecturing, I made little progress. At the time, I worked in divergent fields such as truth and reconciliation processes, democratic transitions, nation building, the history of the liberation struggle in South Africa and public participation – all related, but without an apparent clear link. That was until a colleague reminded me that the missing link was my personal experience of having lived through an apartheid society, having participated in the search for alternatives and having been closely involved during the transition from authoritarian rule to democracy in South Africa. From authoritarian rule to the inauguration of the new constitution in 1996 spanned the 1970s, 1980s and the 1990s. I lived through the trials and tribulations of these eras, he reminded me. In my case, a warehouse of personal and shared experiences accumulated, waiting to be shared a soon as "I", as a tool of qualitative research, grasped it.

My first research proposal envisaged a descriptive study and some analyses of various models used by post-authoritarian governments to deal with a past of severe human rights transgressions. Post-authoritarian rule polities chose various approaches to deal with a past of human rights transgressions. The chosen options included forgive and forget approaches (such as Spain, Namibia and Zimbabwe), truth and reconciliation commissions (Argentina, Chile, South Africa), mixed approaches such as court proceedings and public *fora* to deal with transgressions (Rwanda). Where a change of government did not take place, government-initiated commissions investigated transgressions of human rights and violence (Zimbabwe, Israel, South Africa). There was space to look at comparisons as Hayner (1994: 597ff.) and Bronkhorst's (1995) work in this field proved. A closer look at the South African Truth and Reconciliation Commission (SATRC) and some comparative perspectives on post-conflict reconciliation approaches provided a worthwhile opportunity. The TRC was a case among cases, unique, yet generic, a political process yet replete with human experiences.

There was also the possibility, apart from a comparative angle, to access quantitative data. As a senior researcher at the Human Sciences Research Council (HSRC) of South Africa at the time, I had access to data garnered by annual surveys that included survey questions on the SATRC. This data would have enabled me to mine quantitative data but limitations arose. The SATRC-related questions fell away in the next national survey due to changing research priorities at the HSRC. Possibilities for a quantitative approach fell away. It was time to reflect and reconsider my personal research path.

At the time, I wrote various articles for journals and other media on the SATRC and in some of them compared the SATRC with other truth commissions, such as those in Argentina, Chile, and later Rwanda (in all these cases the military and their abuse of power stood central). I also looked at government-initiated commissions in the aftermath of state violence, i.e. the investigative commissions in Israel after Sabra and Chatila, the Goldstone Commission in South Africa, (failed) commissions in Uganda and Zimbabwe and later on the Oputa Report in Nigeria. My study could gain from an earlier focus on the SATRC and civil–military relations (CMR) in pre- and post-apartheid South Africa. However, since South Africa was not the only country to experience and deal afterwards with human rights transgressions, a comparative element (in contemporary qualitative research referred to as broader casing) entered the picture.

The discussion on the missing link above has relevance. The study leader I had at the time had apparently lost interest (just another reminder that the one important choice in successfully completing a doctoral degree – or for that matter a master's degree – is choosing the right study leader). I switched to another study leader. Together with an already proven co-supervisor, things picked up speed. Here the personal narrative and those of others enter. Auto-ethnography as one qualitative research approach does touch slices of life. It is up-close and personal and tends to become multi-layered.

In opting for auto-ethnograpy some choose for a post-modernist approach. I made a choice not to work in a post-modern paradigm. Violence and war are not relative. Socio-political conflicts are real, not imagined. With apologies to Derrida, endlessly peeling an onion would not bring insights that would benefit future peoples living through the same or similar experiences in the aftermath of organized state violence. Charles Moscos aptly argued that we have reached the stage of a post-modern military (Ferreira 2011: 5). However, none of the countries in my study that saw human rights abuses found themselves in the league of post-modern militaries. They were modern or late modern in composition, command, control and application of their armed forces, even when used against their own citizenry.

The study I undertook focussed on the link between the SATRC and its potential effect on civilian control in a post-apartheid democracy. Within a broader casing, it touched on other cases and the challenges of democratization. Other narratives and personal experiences stepped into the study as the research process unfolded.

Experience, the (personal) narrative and the social context

Personal experience played a central role in the study. This flowed from exposure to militarization in an apartheid society, e.g. the school cadet system and participation as a sixteen-year-old youth in home defence units or territorial militias (known as *Kommando's*/Commandos) and later on as conscript. For South Africans born in the 1960s, as was my case, apartheid was firmly entrenched (South Africa left the British Commonwealth in 1960 after criticism against the imposition of apartheid and declared itself one-sidedly a republic in 1961, after a whites-only referendum). In the United Nations, criticisms increased and South Africa was to see increasing political isolation. In the international arena, South Africa aligned itself closely with other states that had a history of authoritarianism, exclusion and severely restricted democracy, such as Argentina, Chile, Uruguay, Paraguay, Zaire, Malawi, Israel and Taiwan. The white minority, led on by their National Party elite and the secret Afrikaner Broederbond, doggishly clung to power in a society where racial domination kept refining itself with observable negative social effects such as oppression and brutality.

Personal experiences feed into auto-ethnography. For example, at secondary schools, we had a cadet system where boys (and girls, voluntarily) took part in drill and shooting exercises. Approximately 500 white schools countrywide participated in the system. This served as preparation for military service and, on a hidden curricular level, to instil patriotism and, to an extent, racialism. White boys at secondary schools were annually registered for military service. I matriculated in 1978 (with a cadet rank of sergeant and force number starting with "76", indicating registration in 1976) and started two years of military service (conscription was known in South Africa as *national service* or in Afrikaans as *nasionale diensplig*). Training in the infantry and completing a junior leaders' course, I became a lieutenant (platoon commander) and completed three border stints (operational deployments), two as platoon leader.

Following childhood and conscript experiences and further studies, I became a politically conscious, anti-apartheid activist, witnessed before the SATRC on the role and effect of conscription and provided policy advice on some aspects of the interim constitution during my tenure at the Human Sciences Research Council (HSRC). I became a civilian delegate to the Defence Review Process (DRP) that followed the release of the first white paper on defence entitled *Defence in a Democracy* (May 1996), under the new Constitution. Personal involvement as participant, participant-observer and observer-participant provided data and reflected-upon experience. Interaction with the civil and military/security community added to the collage. Thus from 1967, when I entered school, until 1997/1998, when the defence review process going on in South Africa (as well as the SATRC legislation promulgated in 1995), I had 30 years of personal experience as observer, participant observer and observer participant; certainly a basis for an auto-ethnographic study.

I intended to tell my own story and the story of other individuals in similar contexts in such a way that it would generate greater understanding of developments in the chosen field. To do so placed me firmly within an auto-ethnographic approach. "Telling *a* story" (not "*the* story") is not a matter of solipsism or unqualified polyphony. The personal narrative remains grounded in the immersion of the researcher as a human tool in (shared) experience, scholarly and other written materials, including archival material. The personal narrative includes lived experiences and the written, scholarly and reflected-upon data, closely relating slices of lives of real subjects, the participants to the study in this case. A comparative element – or broader casing – entered the narrative or, by now, multi-layered narrative. The discussion with the colleague that pointed out personal involvement and experience came full circle.

I aimed to provide a "feeling" for the participants and my experiences but remained aware of overstating individual experiences or uncritically exporting or transferring advice from one context to another. Added to this I aimed to provide through this exploration some policy pointers and recommendations for others in transition from authoritarian and/or oppressive rule to democracy, and to outline possible implications for civilian control over the military. After all, the institutionalization and nurturing of an attitude accepting civil control and oversight over the military, involves human subjects (the "I" and the "We"), includes social action against oppression and intertwines with the ideal of democracy, whether from a liberal, social democratic or more radical perspective. Here the personal experiences of various interviewees (research participants) provided valuable data. Drawing on multi-layers, I could contribute insights to social-scientific knowledge in local military sociology and prioritize areas for further research without burdening the reader or future researcher with the pretence of objectivity or relativizing the human experience.

As a subtext, I took a cue from Meehan (1988: 14): The "purpose of systematic research is to better life for some people somewhere". There are many reasons for research on particular social realities, problems or phenomena. Research simply for the sake of curiosity had little value for me. One reason for research is to contribute in some practical way. Personal experiences and the socio-political changes taking place in South Africa, together with the conviction to better our situation in the aftermath of authoritarian rule, furthered my commitment to the chosen topic. Social problem solving intertwines with issues such as guarantees for human rights through civil control over the military and civil–military relations (CMR) to enable security governance. Auto-ethnography could prove helpful here.

Experience, criticism and people

An underlying aim incorporated and garnered over years is that I chose, where possible, not to exclude laypeople who contributed to my knowledge and shared experiences. In the process of finding my way to assist in bettering society there were many sources. Insights gained, experiences lived and knowledge gained and shared do not belong to one person, but many, and does not only reside with university degrees or even school education. In knowledge, one stands on the shoulders of others' experiences and foresight and, frequently, we stand amid our own and fellow travellers' hindsight and foresight. Consciously I also exposed myself to non-academic people, practitioners and those from various societies who suffered oppression. But exposure to, and openness to others is not all.

I consulted masses of literature over the years. As my focus became clearer, I had a warehouse of literature available for the literature review. At the same time I had to read closely about auto-ethnography. While I was versed in qualitative approaches in general, interviewing individual participants and conducting focus groups, auto-ethnography was a different genre. Sufficient reading was as vital as consistent interaction with the research participants.

Critically reflecting on my approach and in discussion with others, I started drafting a schedule for interviews with possible participants. I was fortunate to have a study leader who was an expert in qualitative research and auto-ethnography. I had ample interaction with academics, practitioners and others who had participated in or lived through our era of oppression and transition. We had regular discussion groups between students of auto-ethnography facilitated by the study leader. In the *broader casing*, other countries and people that had had similar experiences became important. A visiting academic from Uganda for whom I acted as a host, played a vital role. He shared with me his experiences from Uganda and the Great Lakes region. I could test my interview schedule with him and refine it. Fellow students on the research path and peer debriefers (I chose three people) commented on schedule and text, thus facilitating refinement. Allow me to add here that on the research path circumstances evolve and those you interview are different individuals with their own unique experiences. In this sense, an interview schedule is never complete; it evolves during the research process.

The reader will discern that research collaboration takes place on various critical levels. Potter (1996: 109) identifies options regarding the extent to which qualitative researchers may collaborate when gathering evidence:

1 Sharing analyses among researchers, or *horizontal collaboration*;
2 Collaboration of researchers with research participants, or *vertical collaboration*.

In auto-ethnography the same happens. However, it is not only about the collaboration with other researchers. Here it applies pertinently to venturing into the field, the right to entrée and the participants. It also touches the feedback loop between peer-debriefers, participants and myself as a research tool. The focus is not so much on analyses, but the horizontal and vertical meshing of shared experiences. I may add a third level here: collaboration and living with practitioners of a non-academic background and their "lived-through experiences". To add value, I deployed these three categories in garnering data. Sparkes rightly argues that the researcher is an active participant. It is important that the researcher's location of the self is understood (Sparkes 2002: 17). The location of the researcher impacts on the social setting inhabited by the emotional selves of researcher and participants (Sparkes 2002: 17–18) It is here that the need arises for the researcher "to be written into the research through reflection" (Higate and Cameron 2006: 221).

At the same time, I had to balance my act. I report on personal experiences and emotions, but did not take the genre that I deployed to the level where a "heightened consciousness of textual interpretation" succumbs to a "fictional autobiographical ethnography" or a full-blown "mixed genre" that Plummer (2001: 34–35) refers to. Auto-ethnography is not writing a biography or fiction. What I wrote related to the real experiences of real people. Ethical imperatives remain (Preston-Whyte 1990: 239ff.; Guillemin and Gillam 2004: 261, 263–265). I abided by the ethical prescriptions, norms and universal codes laid down by social science research communities, such as not causing harm to research participants, not invading their privacy and not misleading participants. I chose to gather data overtly for auto-ethnography. And I chose not to invent fictional scenarios, which happens in some auto-ethnographies where mixed or blurred genres come into play. I tried to steer clear of blurred genres, primarily because my point of departure was not post-modern. It related to the narrative of the self and others who had lived through real life experiences. Neither did I approach my research covertly but in the open. After all, transparent research is not and should not be an exercise in intelligence gathering or investigative journalism.

Obtaining entrée and reaching agreement with people involved in the research setting represent crucial decisions. I did not plan all interactions. I gained access without a direct request in some cases. At other times, access came at the cost of consistent effort and energy spent. The researcher will experience rejection in some cases and entrée in other cases and must accept it as part of the research process.

Venturing into the field is rarely easy. Snowballing research participants is challenging. To get hold of some participants was difficult. Then the real work starts. To gain the right to entrée requires commitment, understanding and empathy. In some instances entrée can be negotiated. Generally, however, entrée is earned, not a given. Through hard work, but also fortunate coincidences, I met research participants during the course of the study. They were from South Africa and Namibia (these two presented less of a challenge), Rwanda, Uganda, Argentina and Spain – all countries that experienced colonialism, or authoritarian regimes and human rights violations.

Interviews are never complete. Frequently, I went back to the participants and in all cases (fortunately for me!) they were helpful. Interviews were not always easy. One participant had experienced the brunt of the military junta's oppressive measures in Argentina. She lost friends and family members before she left the country for South Africa. There was nothing easy in this interview. If it touched me so deeply, it is difficult to imagine what she went through. Courageously, she did complete the interview and follow-ups. Perhaps her experience touched me more intensely because at the time I had to cope with the death of four close friends Rocky, Ruhr, Elize and my father, in relatively quick succession. Perhaps it was because during activist times I lost colleagues (we called each other comrades) killed by security forces. As for my "previous life" as a conscript I was fortunate enough not to lose close buddies. I never saw direct contact with the enemy, only the bodies of those killed in contact. We did however, train to kill at the time and as an instructor, I, in turn, trained my soldiers to kill.

Particular useful interviews concerning the South African case were with retired officers and two generals from the South African Defence Force and with ex-guerrilla cadres from the African National Congress military wing, Umkhonto we Sizwe. These were less problematic and, to an extent, less emotional. All of us, even from different past experiences, lived during apartheid times and through the transition. Such experiences provided for some shared experiences and for hindsight that enriched the research narrative. Saying this does not mean that at least one interview with an ex-SADF member did not bring up underlying tensions. Yet both of us, in the course of our interaction, succeeded in sharing views that benefited the study and perhaps by doing so demonstrated that while shared visions cannot be achieved, there are ways to further dialogue through a search for accommodation and furthering interactive communication in the aftermath of an era of social alienation.

I soon realized that the schedule for the interviews had other uses. I adapted the schedule for use in soliciting e-mail responses. I had numerous contacts from earlier times and started contacting them. I sent out a few dozen schedules per e-mail, kept the feedback confidential and explained the need and value of this type of study to the potential participants. I ensured that I approached academics, theorists, practitioners and people who had lived intimately through the context(s) in my study. About half of them responded. Some that I contacted indicated that they were too busy, did not feel confident in the area to comment, or suggested others in their place. Where some e-mail participants fell away, I tried to substitute them for others with varying success.

The responses of those who answered the e-mail request were surprising and fresh. The participants shared anecdotes and references to other cases that added much value to my broader casing that now formed part of the study. Furthermore, in at least two cases, I received much welcomed unsolicited materials. A West African respondent provided me with reports that, while in the public domain, were lesser known. He also supplied transcripts of court cases and articles from the local media related to specific cases. Another participant shared notes for briefings and submissions pertaining to specific civil–military-related issues in his country. Another forwarded articles and working papers that she had published earlier, but that did not appear in accredited journals. If she had not done so, I would not have access to them. The e-mail solicitation, which almost came as an afterthought, turned out to be of immense value.

Writing up and knowing when to stop

PhD students can testify about various stumbling blocks along the research path. Once conquered, these stumbling blocks convert themselves into milestones achieved, such as: choosing the topic; refining the topic; submitting your proposal and getting it accepted. The latter, I still feel, is one of the most challenging for any student/candidate who embarks on the research path, no matter which degree. Other haunting challenges include: systematic reading and summarizing; the writing-up phase; working on feedback received from research participants, peer debriefers, the study leaders and fellow students. The final challenge is the painful proofreading/editing process and ensuring a comprehensive source list.

In my case, my first study leader's feedback was not forthcoming and the few times I did receive some feedback it was sketchy and general to the extreme. The study leader's approach simply implied: "You are on the right track; go ahead, we will talk about it later." Regarding the contents, the co-supervisor's feedback was timely and most helpful – in fact highly professional. To change a study leader presents no small obstacle and indeed can be quite a disturbing experience. The process involves people, emotions and bureaucratic challenges. It can be debilitating and have an effect on your motivation and morale. In my case, it certainly did but I had no wish to give up the study after all the energy put into it. With the first study leader replaced, things turned out for the better. Progress, finally!

The next challenge related to my own nature and personal style. The wider one's field of interest, the easier it is to become sidetracked. Sidetracking may be occasionally rewarding and always interesting. However, sidetracks steal valuable time and energy and, in general, do not contribute to progress on a particular study. On the contrary, they can cause unnecessary delays. I had to work hard at turning away from highly interesting but not-really-applicable reading material. Make a note for later and move on!

If the above holds true for materials consulted, another important truth pertains to the writing-up process: know when to stop. For many of us, it is difficult to cut off and walk away. At some point, one has to. In qualitative research and auto-ethnography, we refer to a point of saturation. This means that at some stage the researcher discovers that the incoming information is repetitive and does not contribute any new insights. Doubts may linger when you reach this stage, but trust your instinct. It is time to complete this particular project. You can always pick it up later on as a separate or new project.

Reaching closure is another dimension of this type of research. You have associated closely with research participants, the context, the atmosphere and the emotions – including your own emotions and reflections – and your written text. Here also, one has to leave behind. You cannot carry the study or the participants with you forever. To reach closure, treat the research participants with empathy and respect and a caring attitude, but get closure. To get closure is part of the auto-ethnographic experience, it is part of life, just as medical doctors or psychologists have to, at some stage or the other, distance themselves from their patients or clients.

The audit trial, shortcomings and lessons learnt

Earlier on, I referred to the fact that we do not use terms such as triangulation and verifiability in auto-ethnography. However, we do use terms such as reliability – did you follow throughout the process approaches and methods that provided you with the most reliable data possible? We refer to transferability. Can someone trace your footsteps in doing a similar study?

Have you informed your reader and future researchers about the shortcomings of your study? Did you share with them some of the pitfalls – potential or real? Moreover, did you share with them your own mistakes to serve as lessons for the future? Here, reflections on your research narrative become important.

Reflections and reflective-ness on the research narrative

Qualitative researchers and more so those who deploy an auto-ethnographic approach know very well that reflections play a role (Ellis and Bochner 1996, 2000; Becker 1998; De Marrais 1998; Ezzy 1998; Meneley and Young 2005; Crang and Cook 2007; Gingras 2007). Critical reflection does not only refer to self-reflection and double checking data and yourself. It also involves peer debriefers critical to the field of study and as much feedback from participants as possible.

Subjectivity and reflexivity play a role in auto-ethnography. Reflexivity is the capacity of researchers to acknowledge how their own experiences and context interplay in the processes and findings of the enquiry (Etherington 2006: 31–32). Criticism levelled against auto-ethnography is that of potential bias and the possibility of self-indulgence, solipsism and/or narcissistic tendencies. As reflexive researchers, we need to recognize this moral dilemma in auto-ethnography. During the study, I remained aware of my potential bias and opted to resolve this by reflecting and sharing my thoughts with people who lived through or with similar experiences.

Reflection refers to "a reflectiveness among social researchers about the *implications* for the knowledge of the social world they generate of their methods, values, biases, decisions, and mere presence in the very situations they investigate" (Bryman 2004: 543 – emphasis mine). Accommodating critical insights, and in numerous cases insightful criticism of colleagues, friends and selected individuals to reflect on my work benefited the study; it added value.

The reflexive researcher remains aware that subjectivity is not an end in itself, but that the researcher is a filter, a heuristic tool (Etherington 2006: 125). She/he is aware that the intentions and choices in the research process involve the being (and becoming) of others. Sharing this with the reader provides a measure of transparency. Some insight on this is found in the late Donna Winslow's statement (quoted by Moelker 2010: 6) that "[it is] about becoming a human being". Winslow's view applies not only to the researcher but also to the participants involved in the process.

The human tale or narrative forms an intrinsic part of the life and history of humans (Burnier 2006). I deployed the auto-ethnographic style, telling the story of the SATRC and selected other TRC-type cases and transitions to democracy intent on installing civil control over the military as much as I related personal experiences before and during the study.

Involvement and human agency touch the material and the practical. The study I chose was aimed at the practical by taking a case study, namely the SATRC, to enlighten a specific demarcated area, namely CMR and civilian oversight of security agencies aimed at sustaining a human rights culture and expanding/deepening the democracy in South Africa. The research process flowed into broader casing and engaging people from other similar, but also different contexts.

In conclusion, I succeeded in answering the rather technical research question through auto-ethnography: TRCs, including the SATRC, can potentially contribute to more robust and healthy civil–military relations and the institutionalization of civil control over the military. However, this is not a given and depends on the planning and management throughout the research process. More importantly, from the beginning, the goal of instituting civil control has to be envisioned and built into the transition process. Even then, a TRC's contribution may not hold guarantees for future stability in civil–military relations because regression to authoritarian rule in new or established democracies remains an ever-present possibility.

I shared the experiences of others from their particular context and emotional being. I could draw on and share unique experiences across political and cultural boundaries. The work shared not only a slice of life, but also slices of lives. The final product shared insights on dealing with similar experiences in other societies: not steps derived from hard modelling or quantitative approaches, but collective wisdom and insights on failed political actions through the experience of others.

In military lingua, we frequently talk about "lessons learnt". On a personal level, I learned lessons: Do not allow setbacks to haunt you forever. Re-evaluate, re-group, re-map if necessary, but do move on. Take the criticisms of peers and colleagues seriously. Learn from them, but also do not allow yourself to wallow in sound criticism to the extent that you do not pick up your worn-out tools and continue the work.

I believe that the study I undertook was trustworthy and transferable (in contemporary qualitative research, the words used for valid and repeatable are *trustworthy* and *transferable*). For obvious reasons, qualitative studies cannot be generalized. If it is trustworthy and transferable, the study may provide clues to a similar approach in another context. It may hold some potential for future research in the field and provide tentative pointers to influence future policy positively. I believe that the pointers provided may have value for other societies in similar circumstances. Perhaps one should add that the answer arrived at may be tentative, but the track explored made it a worthwhile experience in itself.

In planning and executing the study, practical considerations played a part and in some cases, played a part with me as author. Studies can also dictate at stages. After all, the researcher is but a tool in a much larger social process. A study such as this can enrich, bring forth new perspectives, invite others into real life experiences and even provide pointers for the future. In a sense, one may find that you not only process knowledge and experience, but you become part of a process that may add value to society and the future.

Finally, while qualitative research is certainly hard-won knowledge, it is knowledge that contains elements of involvement, which can improve society. If one of the envisaged outcomes of the project is a book, a dissertation, a thesis or an article, the act of authorship is to follow and to conclude in a timely manner. During the course of writing up my PhD, I perhaps allowed for too many sidetracks, which delayed concluding the study.

Conclusion

During every study one experiences tension and anxiety about your research participants, progress and deadlines, as well as your own personal abilities. This experience is not unique. What matters is how you manage it. I tried (and believe with some success) to abide by an old *karate* dictum: *Omuyari* – to care for (each) other. The other, in this case being the participants, their experiences, their society, the narrative and the attempt to contribute with others to insights, sharing data and some suggestions as to ways and means of preventing similar human rights excesses in the future. One has to trust that others will pick up and share such knowledge to better society in terms of civilian oversight of the security forces in our own and other societies in similar dilemmas or post-authoritarian realities.

It is difficult to cut off and walk away. In reflecting on work completed, one feels one should have done more. "The more" I decided to leave for future research...

Appendix

Building blocks to the study

Building Block 1 – Context: Context constitutes many dimensions. Quantitative methods such as social surveys can seldom adequately share a slice of real life and its multiple dimensions and a plurality of experiences. Quantitative research lacks immersion in real-life processes and interactions. Through a researcher's immersion in the human data, the certainties are less, but the experiences are real and he/she at least gains by experiencing and sharing a slice of life through qualitative research techniques like observation, participant observation, interviews and shared experiences.

Building Block 2 – On method(s): Every study has unique features due to organizational characteristics and the personalities and roles of individuals in the society or the institutions within which they live. Case studies benefit from a wide reading of terrain and inhabitants, preferably those who have experienced the circumstances rather than those that come to it with predetermined theoretical mindsets. The levels of interaction, i.e. horizontal, vertical, collaborative (and compartmentalized) are relevant here. Methods cannot be rigid and unchangeable. Methodology evolves on the track. The same applies to one's research design in auto-ethnography. New elements emerge, some amalgamated; others left behind (compare Trafford and Leshem 2008: 90, 93, 101ff; Mouton 2009: 143ff).

Building Block 3 – Literature and people: Qualitative research is about going *out there* and *rites of passage* (by the way, rights of passage, the permission to enter other people's world[s] are also relevant). However, it is also a journey or journeys through literature in relevant state-of-the-art publications. Literature is one important rite of passage towards immersion in data.

"A literature review is a description, critical analysis and evaluation of relevant texts – both current and seminal – that relate to your research topic" (Mouton 2009). Thus, one develops the argument for one's own research on the basis of the literature review (Daymon and Holloway 2002: 35). Literature provides building material. Literature can shape the reader. Equally so, experience and social interaction shape people and the involved researcher. By tracing literature, while keeping myself open for the experience and insights of research participants I stepped progressively into a human narrative of society and the life experiences of people in that context.

The literature review, the choice of design and methodology, subsequent fieldwork and collection of data, analysis, reflection and reporting of the findings are important building blocks in traditional research (Manheim and Rich 1981; Platt 1992: 21, 24ff., 29; Bouma 1996). In auto-ethnography the literature review should not predetermine the track. Studying humans is more complex, and prejudiced reading should not dictate one's research.

Building Block 4 – The case study: While there may be a variety of specific purposes and research questions, the general objective is to develop a full understanding of the case. For the study, I chose the SATRC and the interface with the DRP as my focal point. I did not only rely on scholarly literature but on other literature too (i.e. archival sources, cartoons, propaganda pamphlets, letters, notes, videos, DVDs, art works or extracts from diaries or newspaper comments) and on interaction with people involved in the case bounded within the context.

One of the benefits of the case study was its capacity to explore social processes as they develop and unfold. Case studies prove useful when it is necessary to understand social processes in their environmental context and they are particularly appropriate to explore perceived new processes or activities.

Building Block 5 – The case in broader perspective or broader casing: Frequently, one case calls up another. Or, a case study may prove to be inadequate to answer some of your sought-after questions (Yin 1981: 60ff.). Comprehensive case studies may prove helpful in cross-national comparative research. Here, an understanding of the meaning of concepts, relationships between concepts, meanings attached to particular behaviours and the way behaviours relate is beneficial. At this point, broader casing becomes useful. I opted for a case study, keeping in mind the usefulness of socio-historical comparisons. This assumption enabled me to deal with TRCs in different contexts at different times and to a degree comparable, but with different subjective circumstances, and to link these to people's experiences. However, one case calls up other cases.

There are reasons why I included a comparative element or what the qualitative researchers call "broader casing" for this study. Case studies hold relevance in auto-ethnography on various levels, i.e. a person, a group of persons or a specific context or settings (Yin 1981: 58, 61ff.; Platt 1992: 21ff.; Hartley 1994: 208, 211, 213ff.; Lincoln and Guba 2002: 205; Stake 2002: 435, 437–438).

There are limitations to an exclusive focus on one group or nation. Should one wish to improve the ability to explain, then "one way is to take a comparative approach" (Manheim and Rich 1981: 230). Comparative elements yielded important insights, and constant interaction with practitioners and lay people frequently provided insights beyond a single case.

Building Block 6 – Individuals play a role: The researcher as individual and one of the research tools comes into play. Research of a case or cases cannot be divorced from the process or findings. Nor can the individual divorce her/himself from other agents in the social process, or their contexts, choices and experiences. By *being*, the qualitative researcher (in interaction with others) becomes part of the process, the broader casing.

With this last building block, things come together. Research design, building blocks, relational concepts and agency within evolving contexts mesh. Such fluidity makes qualitative research, especially auto-ethnography, challenging and simultaneously satisfying.

References

Becker, H. S. (1998) *Tricks of the Trade: How to Think About Your Research While You're Doing It*, Chicago: University of Chicago Press.

Bouma, G. D. (1996) *The Research Process*, third edition, Oxford: Oxford University Press.

Bronkhorst, D. (1995) *Truth and Reconciliation: Obstacles and Opportunities for Human Rights*, Amsterdam: Amnesty International.

Bryman, A. (2004) *Social Research Methods*, second edition, Oxford: Oxford University Press.

Burnier, D. (2006) "Encounters with the Self in Social Science Research: A Political Scientist Looks at Autoethnography", *Journal of Contemporary Ethnography*, 35(4): 410–418.

Crang, M. and Cook, I. (2007) *Doing Ethnographies*, New Delhi: Sage.

Czarniawska, B. (2004) *Narratives in Social Science Research*, New Delhi: Sage.

Daymon, C. and Holloway, I. (2002) *Qualitative Research and Marketing Communications*, London: Routledge.

De Marrais, K. B. (ed.) (1998) *Inside Stories: Qualitative Research Reflections*, London: Lawrence Erlbaum.

Denzin, N. K. (2006) "Analytic Auto-ethnography or Déjà Vu All Over Again", *Journal of Contemporary Ethnography*, 35(4): 419–428.

Ellis, C. and Bochner, A. P. (eds.) (1996) *Composing Ethnography: Alternative Forms of Qualitative Writing*, New Delhi: Altamira Press.

Ellis, C. and Bochner, A. P. (2000) "Auto Ethnography, Personal Narrative, Reflexivity: Researcher as Subject", in N. K. Denzin and S. Lincoln (eds) *Handbook of Qualitative Research*, London: Sage.

Etherington, K. (2006) *Becoming a Reflexive Researcher: Using our Selves in Research*, London: Jessica Kingsley.

Ezzy, D. (1998) "Theorizing Narrative Identity: Symbolic Interactionism and Hermeneutics", *The Sociological Quarterly*, 39(2): 239–253.

Ferreira, R. (2011) "The Interdisciplinary Nature of Military Studies: A Sociological Perspective and South African Application", Professorial Inaugural Lecture, University of South Africa (Unisa), Pretoria, 26 July.

Garrat, D. (2003) *My Qualitative Dissertation Journey: Researching Against the Rules*, Cresskill: Hampton Press.

Gingras, J. (2007) "This Could Be: The Possibility of Autoethnographic Fiction", *International Congress Qualitative Enquiry*, Working Paper 3, Canada.

Guillemin, M. and Gillam, L. (2004) "Ethics, Reflexivity, and 'Ethically Important Moments'", *Qualitative Inquiry*, 10(2): 261–280.

Hanna, T. (1970) *Bodies in Revolt: A Primer in Somatic Thinking*, New York: Delta.

Hartley, J. F. (1994) "Case Studies in Organizational Research", in C. Cassell and G. Symon (eds) *Qualitative Methods in Organizational Research: A Practical Analysis*, Thousand Oaks, CA: Sage.

Hayner, P. B. (1994) "Fifteen Truth Commissions, 1974–1994: A Comparative Study", *Human Rights Quarterly*, 16: 597–655.

Higate, P. and Camron, A. (2006) "Reflexivity and Researching the Military", *Armed Forces & Society*, 32(2): 219–233.

Liebenberg, I. (2008) *Truth and Reconciliation Processes and Civil-Military Relations: A Qualitative Exploration*, unpublished DLitt et Phil dissertation, Pretoria: University of South Africa (Unisa).

Lincoln, Y. S. and Guba, E. G. (2002) "Judging the Quality of Case Study Reports", in M. Huberman and M. B. Miles (eds) *The Qualitative Researcher's Companion*, New Delhi: Sage.

Luijpen, W. (1980) *Inleiding tot de Existentiële Fenomenologie*, Utrecht: Uitgeverij Het Spectrum.
Manheim, J. B. and Rich, R. C. (1981) *Empirical Political Analysis: Research Methods in Political Science*, Englewood Cliffs, NJ: Prentice-Hall.
Meehan, E. J. (1988) *The Thinking Game: A Guide to Effective Study*, Chatham: Chatham House.
Meneley, A. and Young, D. J. (eds) (2005) *Auto-Ethnographies: The Anthropology of Academic Practices*, Ontario: Broadview Press.
Mills, C. W. (1972) "The Sociological Imagination: The Promise", in M. L. Medley and J. E. Conyers (eds) *Sociology for the Seventies: A Contemporary Perspective*, Sydney: Wiley.
Moelker, R. (2010) Magical Dragonfly: In Memory of Donna Winslow, *ISA: Research Committee 01. Newsletter*, Accessed from www.ucm.es/info/isa.
Mouton (2009) *How to Succeed in your Master's and Doctoral Studies*, Pretoria: Van Schaik Publishers.
Peperzak, A. (1977) *Vrijheid: Wijsgerige Antropologie*, Baarn: Basisboeken.
Platt, J. (1992) "Cases of Cases ... of Cases", in C. C. Ragin and H. S. Becker (eds) *What is a Case? Exploring the Foundations of Social Inquiry*, Cambridge: Cambridge University Press.
Plummer, K. (2001) *Documents of Life2: An Introduction to a Critical Humanism*, Thousand Oaks, CA: Sage.
Potter, W. J. (1996) *An Analysis of Thinking and Research about Qualitative Methods*, New Jersey: Lawrence Erlbaum.
Preston-Whyte, E. (1990) "Research Ethics in Social Sciences", in J. Mouton and D. Joubert (eds) *Knowledge and Method in the Human Sciences*, Pretoria: Human Sciences Research Council.
Schwandt, T. A. (2001) *Dictionary of Qualitative Inquiry*, second edition, Thousand Oaks, CA: Sage.
Sparkes, A. C. (2002) *Telling Tales in Sport and Physical Activity: A Qualitative Journey*, Leeds: Human Kinetics.
Stake, R. E. (2000) "Case Study", in N. K. Denzin and Y. S. Lincoln (eds) *Handbook of Qualitative Research*, second edition, Thousand Oaks, CA: Sage.
Trafford, V. and Leshem, S. (2008) *Stepping Stones to Achieve Your Doctorate*, Berkshire: Mc-Graw Hill.
Yin, R. K. (1981) "The Case Study Crisis: Some Answers", *Administrative Science Quarterly*, 26: 58–65.

5 Side effects of the chain of command on anthropological research

The Brazilian army

Piero C. Leirner

Introduction

This chapter[1] will analyse relations of control between anthropologists and the military, and some of their effects on ethnographic production. Although, in principle, these relations could be discussed through their operation under two possible schemes – a military anthropology (that "belongs" to the military institutions) and an anthropology of the military (relative to the military personnel)[2] – the idea here is to demonstrate how relations based on the attempt to establish control, between the military and anthropologists, follow a set primary direction. This course can be established in an obligatory manner (when anthropologists work for the military) or indirectly, when anthropologists try to observe soldiers[3] and suffer the "side effects" of their ethnographies.

The discussion begins through observations made during my own fieldwork – systematically conducted between 1992 and 1997, and more sparsely between 2001 and 2010 – but also through the research conducted by my students, which in some ways required my "reinsertion" into the field. Through these experiences, my main objective is to describe the effects of "direct contact" between anthropologists and army officers, resulting from the explicit ethnographic study of the military.

Although this chapter is mainly ethnographic in character, it is also relevant to the recent and academically polemical, direct and normative engagement of anthropologists and anthropological notions – such as "culture", "ethnography" and "alterity", to name the three most important terms – in the United States' military effort in the Middle East (see the criticism by a group of anthropologists against the militarisation of anthropology in the USA: Network... 2009). Although relations between anthropologists and military institutions are hardly new, since 2006 news reports in the USA have captured a new mode of anthropological work: the direct use of the discipline's techniques and knowledge in combat zones.[4] This project is propelled by the idea that it could increase the efficiency of combat units in locations that are the focus of insurgence; the first test was in Afghanistan and the success of the enterprise motivated the North

American State Department to propagate the formula, by proposing that in future there should be at least one anthropologist working in each battalion. This situation raises (at least) a consideration of the limitations of carrying out research with the military.

From where can we begin to consider the problem of writing ethnographies about agents of the State, and more specifically the military? As much as this field has grown recently, it is still relatively small and thus does not yet have a "secure protocol" to at least guide the researcher at the beginning of an ethnographic investment. Whereas, in most cases, anthropologists observe the "implicit rules" of a society, one of the problems of this type of object is that these groups (and the military especially) have explicit rules and protocols from the start, which serve as parameters for their "own" conduct as much as for what should be the "conduct of others". Not that other individuals, groups and societies do not have these, on the contrary, but in this case, the "culture shock" (in Wagner's terms, 1981: 6–13)[5] is not only carried through with elements from "the anthropologist's own culture". These protocols are also made explicit in manuals on conduct, etiquette, classification and planning of the "elementary forms or categories" that define the existence of these groups. Everything is conducted *as if* what the anthropologist normally searches for *implicitly*, among his natives is *explicitly* available; or rather, that practice is somehow given in theory. Thus, we find ourselves forced to "ethnographically invent" elements that could be obvious in other realities, but which are not in this setting.

One of these elements, which will be specifically explored in this chapter, is the inversion of the flux of information between anthropologists and the military: Who questions who? Who is the informant? This will lead us to a series of inversions in ethnographic practice with military, in relation to other more "conventional" ethnographies (for lack of a better term at this time). If the parameters of what we are used to are given in the sense of "anthropologist – object – anthropologist", here, we are faced with a situation that is more "object – anthropologist – object", while trying to convert this equation into "anthropologist – [object – anthropologist – object] – anthropologist". It is a matter of thinking through practical problems in function of theoretical elements often raised by anthropology.

Before proceeding, it is necessary to state that the Brazilian military is very attentive to the academic world and that sooner or later someone would have the brilliant idea of engaging anthropology into the field of war, something that has already been done on a large scale in the USA (Price 1998, 2000). In the latter case, the military expanded its terrain of control by adopting the services of anthropologists themselves, whether by forging anthropologists (that is, sending agents to the academy), or by converting anthropologists into soldiers (that is, annexing the graduate anthropologist to uniform, through mechanisms of conversion that turn the civilian anthropologist into a military anthropologist). This is a new field of relations, far beyond the co-opting of civilians to work *with* the military; it is now a matter of working *as* the military (Price 2002; Gonzáles 2007). In either case, this novelty is not so very frightening, as I believe that *those that have carried out ethnographies with and about the military, have also in a certain way been incited to do it as and for the military*. At least, that was my own case.

Before entering into my specific case, some parentheses are required. It is interesting to consider that the army's co-opting of academics to convert the "culture invented" by anthropologists (in Wagner's terms, 1981) into war reconnaissance, is not always visible to the naked eye and is almost always established in very subtle ways. While this is commonly the case in an anthropology of the Empire or the Nation-State, it took a long time for this to be spoken of in the discipline. We certainly cannot say that this co-optive movement is supported by anthropological bad faith. In a certain way, the extension of State apparatus goes much beyond the organisational capacity of anthropologists in practical terms. One could even say that sometimes co-optation takes place through relational "torsion" devices, such as intermediary agencies, false foundations and sources of funds apparently unconnected to the armed forces. Close parentheses.

In this line, the case that will be recounted here is the result of a crossroad experienced during fieldwork. One point to advance is that from the beginning of the research there had always been a military attempt to co-opt me into "working with/for/as" them. Another point to advance is that "the other side" did not like to be treated as subjects being "researched". Although a hostile position was never reached, there was a strange sentiment in relation to "someone who looks after Amerindians trying to understand us".

The case begins in 1992, but this story continues until today. This chapter will attempt to show how certain ethnographic "side effects" revealed a perception of how the army operates its social life, as well as about the modalities of its projection onto a "field or system of war". Above all, it deals with the idea that the military's attempts at having absolute control over the anthropologist had the effect of introducing a "culture shock" that produced an unusual ethnographic result. In some way, the entire series of prescriptions that military life establishes as its daily routine, produced effects *of* and *for the* research, *on* and *of the* researcher.

Chain of command: initial diagnosis

My first contact with the military was in 1992, through a contact[6] organised by my Master's supervisor, at the time. I arrived at the Army's Command and Joint-of-Staff School (*Escola de Comando e Estado-Maior do Exército*, ECEME), situated in Rio de Janeiro, with a notebook and a "project" – written as a letter of intentions, in which I punctually summarised a programme of research that, above all, requested my deployment to an Amazonian frontier platoon, probably near the town of São Gabriel da Cachoeira (Amazonas State), for a period of at least six months. My first reception was driven by reciprocity – I submitted the project, telephone numbers, addresses, and in return I was given a textbook,[7] which I was to "study and then present a summary": as one officer told me "this is what we do here".

This became a routine for more than two years. During intermittent periods, I spent my days attending activities in the same location and seeing the chances of an authorisation to leave and research the platoons become increasingly less feasible. At various moments, I understood that there was a subliminal message saying that the "place of a university student" was in that space – which was by all appearances what the military world had that most closely resembled the civilian university. More than that, it was assumed that it was there that (to a certain degree) I could come close to "what a soldier thinks", and although I did not perceive this at the time, it appeared to have its desired effect on me, for at a certain point there the insistence grew for me to leave the university and join the army. At the time, I was about 25 years old and this was still a possibility; sometimes I could feel the officers' frustration at my refusal, which was generously understood as something like "my role here is more important" – a phrase that actually replicated various discourses heard before, which claimed that I could have a "fundamental role in strengthening ties between institutions that had the mission to build a project for Brazil".

The relation between these military and myself was elaborated through their systematic attempts to establish a policy of "strengthening ties" with what they understood to be "the university". Note here that my own institution was spoken of in the singular, a fact that was interpreted by myself, at the time, as a reverse understanding of their own corporation; that is, as a type of "*paisano*" (a deprecatory term and category with which soldiers define civilians, cf. Castro 1990) replica of the armed forces, that supposedly guarded two principal symmetrical properties – hierarchy and discipline (Leirner 1997). I understood that at that moment, I was entering the grey zone that follows the canonic stages of the relationship between researcher and natives – which includes "culture shock", the exchange of pleasantries, mutual attempts at reciprocal understanding, the symbolic stabilisation of the arrangement that served to process the associations executed by them and myself, and finally, the control of this learning process and its objectification as "culture": the invention of a controlled "object" (Wagner 1981: 44ff.) – and which allowed me to re-examine the relations experienced in the field in ethnographic terms.

At the time, the thought that the military, like ethnographers, might "invent" a culture (ibid.) and that all of these events could be treated as "ethnographic facts", did not cross my mind. I supposed that this was a type of preliminary phase and so I gave the notion of "strengthening ties" little thought, beyond the need to continue this policy as a means to acquire the eventual authorisation to conduct actual fieldwork. What they called "alliance", "ties" and "policy" certainly were not the same things that these concepts meant to me, which, in the best possible scenario, were concepts saturated with a series of common-sense notions, or ideas taken from political science, sociology or even from an anthropology that was very distant from this "object".

But if these concepts suffer a kind of "creative torsion" or "invention" in the military mind, my place here is to reanalyse these and retrace them from another perspective. Thus, in this article, away from urgent fieldwork requirements, I can reconsider the meaning of ideas of "politics" and "alliance" between the military and the "university". On the one hand, these notions should be encompassed by the grammar of war, this object's reason for existence; on the other, they can be approximated to a language of war that is familiar through broader anthropological literature: as such, these relations can be seen to reflect, for example, the permanent tension that governs the unstable alliance between brothers-in-law, which is resumed through "pacifying" exchange.[8] Strangely enough, a retired general (and ex-minister) once commented something like: "So, how are you going at the university? Still a bunch of communists?", which sounded as if it came from someone who one has not seen for a long time, and who thus uses cordially suspicious words. "Faithful enemies", to borrow Fausto's (2001) idea.

If these notions are indeed indexed by war, the perception of "the university" as a hierarchical and disciplined mirror-reflection becomes clear. The university would be seen as an army; knowledge as discipline; science as strategy; anthropology as espionage; the ethnographer (in the attack) as a double agent, that is, both an informant and propagator of ideas. And, thus, politics can also be conceived as a "continuation of war by other means", as Foucault (1999) suggests in his inversion of Clausewitz's famous saying.[9] It is a matter of taking the idea that "we are at war" seriously: where war is no longer a phenomenon thought through battles or as "such and such a war" that was won or lost. In the native understanding, war is a potential state at every moment, which is currently being deterred: as the officers often insisted, "if you do not perceive the war, it is because we are deterring the enemy".

This is not simply a case of reproducing the native argument. While looking for a notion that could encapsulate this argument, it became clear that war could not be understood as a concrete fact, but should rather be seen as a relation, and this is a point that we are more used to in anthropology; war as a type of social relation between reciprocal enemies. In the end, there is no way in which this native notion would not affect my own concepts and, in this line, we can attempt to identify some others: "alliance", for example, or "exchange". In anthropological literature, these concepts have already been taken as modes of war or at least in relation to it (Lévi-Strauss 1976 [1942]; Clastres 1980; Fausto 2001). In the field, "alliance" was a task from the "strengthening of ties" agenda; the "exchange" between institutions, and between the ethnographer and his subjects, encompassed by hierarchy and discipline – the effect of a command chain. These relations, which were directed by a military routine that is enveloped by war, were also somehow transferred onto the ethnographer. In this case, the ethnography came to be a logical extension of war – a relation, military style.

This does not just refer to events that took place during the first few moments of contact: discipline (and punish). These events take place even today, with students that I supervise and who have decided to study this theme. The ethnographer is studied from the minute he enters the military unit; they know "who he is", how and why he is there; someone is expecting him; someone takes him to the person designated to receive him; and that person will say: this or that interests you and this is what you will do. "This is what should be seen." Today I notice that this is a consequence of something that these natives always said, "the soldier thinks prospectively", he must anticipate the unexpected. It is true that to anticipate the unexpected is not a soldier's privilege in cultural terms, as Sahlins suggested the Hawaiians did it and many others still do (Sahlins 1990, 2008). But few do it as a conscious exercise that transforms cultural categories in action.

As a stranger the ethnographer has to be investigated. How does this begin? In a first contact, the initial step is to have an official letter from the ethnographer's institution, which passes through the hands of supervisor, head, department and university. But this may not be enough. Requests are made so that the ethnographer's immediate superiors – supervisor, departmental or unit head – somehow indicate that they are involved in the process themselves. This is the first sign of the hierarchy's commitment to the ethnographer, that the "university's chain of command" (in the sense of how they understand the university's hierarchical structure, *as if* it has a chain of command) can be called upon if something goes wrong. If the ethnographer changes the military unit that he wants to study, all of these steps must be repeated: once again, a new letter of intentions, stamped by the university, will be submitted. If the ethnographer changes, even within the same military unit, this step is also necessary, and more: if the same ethnographer wishes to study the same unit again, after some time away, this step must be repeated.

This procedure is also an effect of the chain of command. Although the common-sense perception of military hierarchy suggests a "stratified pyramid", what takes place is a much more detailed and complex composition: each individual appears in a singular place in the chain, two people will never be in the same position, there is always someone who commands and someone who obeys immediately "before" and "after" each person. When the chain of command "moves" – for example, during a period of promotion – individuals move together, by changing ranks or posts (Leirner 1997). Thus, the effect of this movement for the ethnographer is to restart the relation from scratch, because as the chain is remade, the ethnographer also "stops existing" in his previous dimensions for the particular section that he tries to approach. It is worth noting that I went through this situation myself on various occasions, as well as going through this again through my students (and they through me in an interminable and, just in passing, unnerving cycle: and once again, the "side effects" appear).

This situation indicates a peculiar relation between individual and the collective group, which is largely unstudied in sociological and anthropological literature. Everything operates as if collective determinations simply encompassed the individual,[10] but notably this hierarchy has specialised itself so much that it reproduces itself individually, and thus appears as an "individualist hierarchy". One of the effects for the ethnographer is that he is perceived as a "representative" of his institution, while the latter must also pass through the approval of its own "chain of command".

One of the most extraordinary things heard repeatedly in the field was the question of whether someone was a "friend or enemy of the army", that "such and such a person was a friend of the army", or that another "had been a friend, but had become an enemy of the army". At first I thought that it was just an expression, but after some time I saw that the binary friend/enemy was absolutely central as a native category. Its importance is based above all on the quantity of dimensions that it is able to articulate: countries, armies, commanders, politicians and simple ethnographers can be friends or enemies of the army. In a certain way, this indistinguishable scale can be understood as one of the chain of command's principal effects, wherein it is able to include the ethnographer as part of a foreign army. But above all, what it reveals is that the category "friend/enemy" is imbricated in the chain of command; that is, that it can be seen as an "extension of war by other means". Some ethnographic information is required here so that the extensive quality of this proposition can be understood.

The observation of the daily military routine in action can provoke questions such as "how does the way in which one sits at the table have anything to do with how one combats?" This type of question returns us to the idea of prospection, as well as to the codification of military life; with such a perspective on military recruits, which aspects of the daily routine mark their socialisation? Celso Castro (1990), whose research was conducted through an ethnography of the Agulhas Negras Military Academy (AMAN), shows that from the very first moment of his four years at the boarding school, the recruit is submitted to a battery of expiatory rituals, physical training and the constant repetition of mnemonic exercises, whose function seems to be "naturalised" inculcation or "memorisation" of military principles.[11] These mechanisms seem to have a dual aim: (a) to stimulate constant desertion among the cadets, so that those who persevere incorporate the notion that they have a "natural vocation" for military life; and (b) forging the construction of a new person, whose new identity is recognised through the notion of belonging to an "internal world".

Such recognition takes place by constantly updating principles of reality in relation to the hierarchy (Leirner 1997) and through distinctly holistic characteristics (cf. Dumont 1992). This is concretely seen by the natives through their engagement with discipline. Differently from "us" (*paisanos* to them, but mainly "us" from the university, the officers' principal comparative counterpoint), with the diverse disciplines that we pass through as *part* of our lives, soldiers have an entire prescriptive regime condensed into a unique source of "military capital", known as *discipline*. Thus, if our etiquette can be disassociated from our "intellectual disciplines" (I can be a brilliant anthropologist with horrendous manners, or mediocre but decorous), etiquette cannot be disassociated from military discipline. Military regulations foresee rigour as much for combat formation, as for a parade and how to enter an elevator.

In the barrack, everything passes through a prescriptive regime, from sitting at a table to walking in the corridor, or speaking, greeting a colleague, taking part in a funeral, writing a memorandum, entering a vehicle and to combat. Orders and rules that are fixed by the chain of command must be followed, and these are generally available to military personnel through disciplinary and etiquette regulations. For example, in the Brazilian army, sitting at a table abides by the following rule: the first superior officer (let us call him "ego") sits in the centre, subsequently, others will sit immediately to the right and left of ego following the hierarchic order until all of the seats have been taken. In a basic infantry assault operations manual, in principle, the same rule should be followed in relation to combat lines, always supposing that one of the maximum objectives is to preserve the chain of command.

As a counter example, the contemplation of a break in the chain of command generates a type of "horror of incest" among the military. In one case observed in the field, a story was told about the relationship of an officer with a subordinate placed several levels below his "hierarchic circle". The native classification for this type of relationship is "hierarchic promiscuity", which is associated with a series of taboos and is taken as one of the worst horrors that can take place in military life. This strong term evoked a series of restrictions and rules that mark the game of alliance, and thus makes one consider the chain of command as an aspect of this. War impresses its meaning on the chain of command. It is, after all, ordering the world, and the idea of "promiscuity" could have its origins in the intention to evoke a general principle of classification, something similar to "each in his place", which obviously includes the ethnographer and his intentions (Leirner 1997). Another message was made very clear to me on another occasion: "Piero, hot soup should be eaten from the edges."[12] It is always necessary to be attentive to the place that one occupies in the chain and to know which exchanges are possible and which are not.

As observed by Castro (1990), it is curious to note that we are speaking of a reality that clearly states that there is a separation between "us" and the "external world", and that these two worlds occupy different places in a hierarchy with its foundations in war. If in our own world – supposedly "scientific", or at least a world where values and culture are "invented" by anthropologists – the *memorisation* of ideas is a lesser form of its conceptualisation, and where double understandings, paradoxes and the coexistence of antagonistic paradigms are worth more, in the military world the constant and repetitive marking of reality suggests that terms and concepts require the unification of word and action. Evidently, this form of reading (and producing) reality (as any other) is also subject to ambiguities from our point of view; nevertheless, for the military it is a matter of (tentatively) always converging for a unified vision.[13] Thus, "attention!" means the immediately corresponding corporeal posture: the interval between order and consummation of the act, between command and obedience is reduced to a minimum (completely, ideally).[14] One thus notes that the ethnographer slowly has to indicate a vector in the direction of this register, if he or she wants to continue researching the military. Actually, this is the case for all the researchers that appear among the military now and then. For example, I was told of situations in which enthusiastic International Relations students "militarised" themselves, by marching, intoning the voice in a peculiar way and singing the national anthem in an almost martial style.

This is one of the effects of mechanisms that attempt to "minimise" the civilian characteristics in the *individual* – be he the Armed Forces recruit or he who wants to cohabit with soldiers – in aiming at a supposedly natural "military essence", while at the same time trying to fill that space with something[15] (and thus we are all potential soldiers and, according to the military, because of this "while the human is human, there will be war"). The social engineering that executes this is based, above all, in a ritualised daily life, entirely marked by the repetitive ordering of reality. This can be traced in the constant graphic representations of timetables and modes of conduct; the automatically recognised mechanisms of action such as orders, corporeal postures and etiquette; the recognition of symbols and notes, such as the emblems and signs that are stamped on uniforms; and, finally, through terminology characterised by the employment of a language that is encoded through acronyms and native terms (cf. Leirner 2008).

External signs that are produced in events such as, for example (and to get back to the subject), during an ethnographic study about them, are also lived and codified. The codification and ritualisation of the ethnographer's life is in this sense one of the greatest signs that he or she has entered native life, and that he or she is part of the tribe, whether as friend or foe. To return to the "side effects" of this relation, or as Favret-Saada (2005) suggests its "affectation", many signs can be detected in the ethnographer's transformations: in my case, paranoia, persecutory mania, the constant sensation of being watched. I would frequently notice interference in telephone conversations and became particularly susceptible to stories about secret bases underneath the *Lago Norte* (North Lake) in Brasília, international conspiracies and the permanent impression that the world hung by a thread. For two years, as part of my fieldwork activities, I attempted to memorise the principles of the "art of warfare", whether by exhaustively reading Clausewitz or Sun-Tzu, so that I could see that "life is war and interests", as I was told.

Curiously, my students also had their own "affectations": after spending two weeks in a regiment, one student spent months repeating the timetable to which she had been made to follow, as well as "returning with an organisational obsessiveness, that had previously been absent". Another student, exhausted with the procedures involved in attempts to begin research, decided to accept that he had "been beaten": numerous letters had returned requesting "more details", "more explanations", "referral to another section", and he noticed that this was an endless cycle. In many ways methodical, the "civilian mirror-image" of military discipline, the student lost himself in this limitless temporality of his object's barriers. In other cases, I observed researchers become impotent before the well-documented feeling of ethnographic abandonment on going "back to zero", after the chain of command had shifted and no-one had made this clear. Finally, a few years ago, while discussing differences with researchers from this field ("milicologists", as we are formally called in Brazil), but who do not conduct ethnographies with soldiers, Celso Castro and I came to the conclusion that there really is something implicitly different in the quality of information collected, the moment that one adopts the "*paisano*" (civilian) position, something that is very different to being a "researcher".

I would not want to suppose that these situations are a matter of pure psychology; although they take place at a psychological level, they could only take place as a provocation of the ethnographic context. Thus, it is worth understanding how and why these transferences, projections and affectations were generated, and transform the psychological data into anthropological data. This chapter will, therefore, end by returning to the idea that the type of ethnographic exchange that takes place during fieldwork with military, inverts apparently "traditional" elements of ethnographic research.

Who is the informant?

When we normally read in monographs, the (often summary) indications of the relation established between the ethnographer and members of the group being studied, it is common for the anthropologist to assume the position of relative, friend, confidant, or even the complicated political implications expected in the *devolution* of the anthropologist's work to that group; either the group simply does not expect anything in return, or this is omitted.[16] Either way, there are few cases *in which the object is more interested in knowing about the ethnographer than being known* by him.

Perhaps this "detail" makes all the difference. Especially in relation to the absolute difference between the *meaning* of information when speaking to either active duty or retired officers. During the first years of fieldwork, I systematically noticed that conversations with my active duty informants were more marked by silence and omissions than by information. Retired informants, on the other hand, always claimed to be "permitted" to talk, and filled boxes full of cassettes. Analogously, it was common – as described above – for active officers to "forget" me, which constantly brought my research "back to zero".

A very good example of this took place during a visit to a military unit in 1995, when I noticed that the officer who had invited me actually crouched down and made a swift exit behind some furniture to the unit's car park when he saw me arriving, he then took a vehicle and "went out". Aside from the pathetic side of this situation, seeing that he had insisted on my visit just a day earlier, I concluded that this corresponded to a repetition of elements witnessed on several occasions: unfulfilled invitations, orders "that didn't come through" and/or "lapses" of information. When an invitation was sent, upon arriving at the location I would hear "I don't think this is here, you need to go to…", something which happened to myself and others. One of my students noted that this procedure is actually a common form of resolving requests for authorisation: a letter requesting access to a military unit travelled the São Carlos–São Paulo–Brasília itinerary there and back twice, before being "lost" along the way on the third attempt.

These events have different meanings in the "exchange of words": if active soldiers "receive and don't return", the retireds "give without asking for a return"[17]; but evidently this is just one piece in the game. Without a doubt, the most interesting element is the relational dislocation of *who asks and who responds*. Normally, we call our interlocutors "informants"; in military language this means someone who "works" for them, infiltrated within the enemy lines. On many occasions I felt myself to be in this role: it was common for me to be put through the controversies surrounding my intentions, my career, my research institution, what we (institution, anthropologists) thought of the Amazon, the Amerindians, if we were communists and, principally, what view we had of the army. On many occasions I noticed the use of interrogation techniques, with the same question being intermittently repeated. The same can be said of letters, projects and research intentions: my students and I had to pass through the process of rewriting the same thing several times, "better specify", give more details, etc. As explained above, this was accompanied by tasks and instructions. And when I wrote letters sending my work, I would receive replies with notes: "this is wrong, this is right", etc.

It is notable that military institutions, such as the Brazilian army, try to obtain control of practically everything that is written or said about them. Each military unit in Brazil – and I believe that this is the case in most armies (Leirner 2001) – has a specific section responsible for the collection of local information, where and when terms such as "army" and "military" appear. In the Brazilian case this is carried out by the "MOs" (military organisations) 2nd Section (usually S2 in any army), which collects, organises, reviews and sends all material up the chain of command. For example, as Celso Castro attests (in a personal communication), it is common for the military to read the review for one of his books, prepared by the command's staff in Brasília. In this situation, what matters is that the chain of command has captured the information and that it has been internally processed.

In the case of my own ethnography, after some time and various contacts and attempts by established contacts, the guidance of an (active) officer who would know how to answer to my needs was recommended. For two years I was accompanied by him in the field, and he was also present at academic social science congresses, together with a team of subordinates who would carefully write down all they heard at the round tables and conferences. I came to know later that this subject[18] was what was known as an E-2, or rather, an intelligence officer from a sector of the General-Quarters in Brasília, responsible for detecting representations of the army within the social sciences. An interesting detail is that he had a doctorate in Sociology from a prestigious Brazilian university. In its own way, the army also trains their "inventors of culture" (Wagner 1981) to detect other tribes – in this case, anthropologists and other social scientists.

Even if we consider this movement as part of a broader relation of the "administration of alterity" (Lima 1995), and although much of the anthropologists' "use" lies in learning their methods and (doubly) converting them so that they can work with/as soldiers, it should be remembered that we are confronted with a field that has an enormous quantity of contextual variables. We can be the army's "friends" or "enemies" in a series of circumstances and we can always commute between these two positions. It is true that on many occasions I also experienced confusion in certain situations where "an anthropologist should naturally be the friend of Amerindians, and therefore our enemy"; or even: "we are true Amerindian friends, you can be our friend"; or the more surprising: "it is 'a laugh' being a native, I never thought I would be one...".

All of these positions are therefore relational and contextual. If, as one colonel told me, "the United States were once our friends, they are no longer", we can also be, or not. But as in the United States, from the military perspective, anthropologists – as well as others responsible for a "national project", as it was always put to me – must be inserted into a logic of "strengthening ties" in any way. The current context (post-September 11th), which is founded on the camouflage of the field of war, as well as on the scarcely foreseeable centrifugation of the armed forces, shows that OPSis (*operações psicológicas*, psychological operations) have been one of the principal resources of various machines of war throughout the world. And, without meaning to extrapolate too much on the "side effects" of military ethnographies, I would also ask myself how much these "psychological operations" are not used in tactics for attracting anthropologists and their work.

One of the strange results or effects that I noticed after the end of my first ethnography was the incredible closing-down of previous research sites when embarking on new research. Curiously, another researcher – a political scientist, and also an ex-army officer – that studied places very close to my own and who finished his dissertation at around the same time as myself, also noticed this closing-down. Both of us thought that we were to "blame"; how else to understand that from one minute to the next a simple military school library became restricted to military personnel only? When I finally met a previous informant and asked what had happened, the response was an inconclusive "you know…". Obviously he did not "know"; in fact, he knew through side effects that silence was the response to "my chain of command" – "university" – which needed to be "refrigerated". In a certain way, it is the same "refrigerator" that "cools" documents and requisitions, which induces the researcher to circulate in an inconclusive chain.

After some time, I also felt the "cyclical effects" of things to which I had become accustomed. In an excursion to Amazonia organised by the military, Celso Castro invited me to finally accompany him on a visit to the frontier platoons, travelling through various units over the course of a few days. Suddenly, while in Manaus, on the day we were supposed to embark onto the military aeroplane, our names had mysteriously "disappeared" from the list of passengers. As one officer told us later, while trying to fix the situation at that time of year (December), "one never knows when an aeroplane will go to those parts…". But we had one consolation: an excellent visit to the Centre for Instruction in Jungle Warfare (*Centro de Instrução de Guerra na Selva*) in Manaus, "where many tourists go. There is even a zoo with a beautiful jaguar, which is our mascot here…", he repeated.

Some years went by, at some point during this time we were seemingly taken out of the "refrigerator" that Brazilian researchers had found themselves in for ten years. We could once again go to the military library or retrace the whole itinerary of research. The most incredible aspect of this was that my students, when asking officers about myself or about past research in the institution, are confronted with the vacuum of memory loss – perhaps beneficially, in this case. Today students who do research on this theme, tell me that "no-one" has heard of the ethnographies written during the 1990s. Sometimes, "one or the other has heard of them, but don't really know what it is about", as one student informed me. Nothing personal, today I am sure that this is one more of the "effects of the chain of command". Probably, this is a cycle that will repeat itself again. Probably ... because, as I was told, "war is the field of uncertainty, par excellence".

Notes

1 This chapter was first elaborated from a class given during post-doctoral research at ICS/University of Lisbon. I would like to thank Helena Carreiras, Celso Castro and João de Pina Cabral for their encouragement, and Julia Sauma for the English version of the text. Obviously the errors and mistakes contained within the article are entirely my responsibility.

2 It is worth indicating that the aim is not to cover all of the literature on the sub-area called the "anthropology of war", although in some way this theme appears as part of the relations between the military and anthropologists, as will be shown.

3 Here I'm taking "soldier" in a generic sense for any military personnel, distinguished, thus, of the army rank of "soldier" – probably the lowest grade in the hierarchy of most of the world's armed forces. Indeed, it's a common sense in military circles that the *military persona* is defined as a *soldier* above all other adjectives that can illustrate someone's social condition. Thereby, an army officer is first and foremost a soldier too.

4 The most notable article is published in *The New Yorker*: "Knowing the Enemy: Can Social Scientists Redefine the 'War on Terror'?", by George Packer, 18 December 2006: www.newyorker.com/archive/2006/12/18/061218fa_fact2. See also, for example, the *Newsweek* article "A Gun in One Hand, a Pen in the Other", www.newsweek.com/id/131752. Accessed 24 April 2008.

5 On culture shock, Wagner states:

> This feeling is known to anthropologists as "culture shock". In it the local "culture" first manifests itself to the anthropologist through his own *inadequacy*; against the backdrop of his new surroundings it is he who has become "visible". The situation has some parallels within our own society: the freshman first entering college, the new army recruit, and anyone else who is compelled to live in "new" or alien surroundings, all have had some taste of this kind of "shock". Typically the sufferer is depressed and anxious, he may withdraw into himself, or grasp at any chance to communicate with others. To a degree that we seldom realize, we depend upon the participation of others in our lives, and upon our own participation in the lives of others. Our success and effectiveness as persons is based upon this participation, and upon an ability to maintain a controlling competence in communicating with others. Culture shock is a loss of the self through the loss of these supports. College freshmen and army recruits, who find themselves, after all, in another segment of their own culture, soon establish some control over the situation. For the anthropological fieldworker, however, the problem is both more pressing and more enduring.
>
> (1981: 6–7)

6 Later I came to know that it was a retired general that worked for an institute that co-opted sections of the Brazilian elite to establish a "project for Brazil".

7 The texts had varied content: copies of news articles about Amazonia, texts about Clausewitz and the concept of "centre of gravity" in military strategy, even a selection of phrases and dictums by Sun-Tzu and Patton, Lenin and Mao-Tse-Tung.

8 Although exchange in itself is not "pacifying", it can initiate new tensions and contains in itself a combative potential, as already attested by Lefort (1979) and Bourdieu (1996), among others. An interesting essay that returns to Lefort and Bourdieu's, as well as Sahlins on the problem of politics in reciprocal relation, is Villela (2001: 191–197).

9 Foucault places "Nietzsche's hypothesis" among others used by him to think through political power. "In this hypothesis, political power would have as its function the perpetual reinsertion of this relation of force, through a type of silent war, inserting it in institutions, economic inequalities, in language and even in some bodies. It would thus be the first meaning given to this inversion of Clausewitz's aphorism: politics is war continued by other means" (Foucault 1999: 23). Thus, the series war-politics can be thought of as acting to absorb a certain vocabulary, that is linked to war terminology (tactic, strategy, etc.) by institutions and disciplinary practices.

10 See Dumont's notion of the chain of command's "artificial hierarchies" (1992: 100), which cannot be applied in this case, as we shall see.

11 Which takes us to the notion of "memorising" an idea (or putting facts as pieces of a mnemonic table), transforming memory into a fact of the "heart" (in Portuguese speech, the verb is "decorar", as "learn by heart"), then, transforming word into action. This also seems to be one of the characteristics indicated by Ben-Ari (1998), on Israel's defence forces.

12 "Piero, sopa quente se come pela borda."

13 Let us say that there is a divergence between two soldiers. What matters is that it has to be resolved, this situation cannot remain "stuck" in the chain of command. Thus if one exterior truth or theory opens a divergence in the chain, at the end of it all it is necessary for a single truth to be chosen.

14 Two military maxims illustrate this "spirit": "When the brain does not work, the arm flexes!"; "Only exhaustive training leads to exhaustion…"

15 The objective of the cadets' "basic course" (the first two years of training in AMAN) is thus defined:

> In the 1st and 2nd years at the Military Academy, the future officer's Basic Training takes places. The objectives are to adjust the personality of the cadet to the principles that rule military life, to assure the knowledge that will enable the continuation of their officer training, strengthening the military character, preparing for basic combat, obtaining basic reflexes for the execution of individual combat techniques and tactics, obtaining physical training and developing technical abilities.
>
> (www.aman.ensino.eb.br/pvisaogeral.htm, on 26 July 2006)

16 Although this topic has already received a bit of attention here, most of the problems that the issue claims are beyond the remit of this chapter. I would only note that in the soldiers' case, or at least in related cases, to my knowledge there are few ethnographies that discuss the relation with the so-called "powerful". In relation to soldiers, aside from those already cited – Castro (1990) and Leirner (1997) – see Badaró (2009). For the question of control over ethnographies by elites, see the introduction in Pina-Cabral and Lima (2000).

17 It is impossible not to think of Clastres' famous essay "Exchange and Power: Philosophy of Indigenous Chieftainship" (Clastres 1977). Clearly, these "temporal lapses" in reciprocity can also be considered together: those that speak "after" retiring, "returned" what active soldiers only "heard". If we would like to

continue in this direction (on another occasion), it is possible to think of this reciprocal temporality through the politics of its relation (cf. Bourdieu 1996). For a critique of the problem of "devolution" by the chief in Clastres, see Lanna (2005).

18 Through information provided by a retired officer.

References

Badaró, M. (2009) "Dilemas da antropologia das instituições controvertidas: reflexões a partir de uma investigação etnográfica no Exército argentine", in C. Castro and P. Leirner (eds) *Antropologia dos militares: reflexões sobre pesquisas de campo*, Rio de Janeiro: FGV.

Ben-Ari, E. (1998) *Mastering Soldiers: Conflict, Emotions and the Enemy in an Israeli Military Unit*, Oxford: Berghahn.

Bourdieu, P. (1996) "Marginália: notas adicionais sobre o dom", *Mana*, 2/2. Rio de Janeiro: Contra-Capa.

Castro, C. (1990) *O Espírito Militar: um antropólogo na caserna*, Rio de Janeiro: Jorge Zahar Ed.

Clastres, P. (1977) *A Sociedade Contra o Estado*, São Paulo: Brasiliense.

Clastres, P. (1980) *Arqueologia da Violência*, São Paulo: Brasiliense.

Dumont, L. (1992) *Homo-Hierarchicus*, São Paulo: Edusp.

Fausto, C. (2001) *Inimigos Fiéis: história, Guerra e xamanismo na Amazônia*, São Paulo: Edusp.

Favret-Saada, J. (2005) "Ser Afetado", *Cadernos de Campo*, 13. São Paulo: USP.

Foucault, M. (1999) *Em Defesa da Sociedade*, São Paulo: Martins Fontes.

Gonzáles, R. (2007) "Towards Mercenary Anthropology?", *Anthropology Today*, 23–3: 14–19.

Lanna, M. (2005) "As Sociedades contra o Estado existem? Reciprocidade e poder em Pierre Clastres", *Mana – Estudos de Antropologia Social*, Rio de Janeiro, 2: 419–448.

Lefort, C. (1979) *As Formas da História*, São Paulo: Brasiliense.

Leirner, P. (1997) *Meia-Volta, Volver: um estudo antropológico sobre a hierarquia militar*, Rio de Janeiro: FGV.

Leirner, P. (2001) *O Sistema da Guerra*, PhD Thesis. Dept. of Anthropology, São Paulo: USP.

Leirner, P. (2008) "Sobre 'Nomes de Guerra': classificação e terminologia militares", *Etnográfica*, 12(1). Lisboa: ICS/Univ. de Lisboa.

Lévi-Strauss, C. (1976) [1942] "Guerra e Comércio entre os Índios da América do Sul", in E. Shaden, *Leituras de Etnologia Brasileira*, São Paulo: Companhia Editora Nacional.

Lima, A. (1995) *Um Grande Cerco de Paz: poder tutelar, indianidade e formação do Estado no Brasil*, Petrópolis: Vozes.

Network of Concerned Anthropologists (2009) *The Counter-Counterinsurgency Manual, or Notes on Demilitarizing American Society*, Chicago: Prickly Paradigm Press.

Pina-Cabral, J. and Lima, M.A.P. (2000) *Elites: Choice. Leadership and Succession*, Oxford: Berg Publishers.

Price, D. (1998) "Gregory Bateson and the OSS: World War II and Bateson's Assessment of Applied Anthropology", *Human Organization*, 57(4): 379–384.

Price, D. (2000) "Anthropologists as Spies", *The Nation*, November 20: 24–27.
Price, D. (2002) "Past Wars, Present Dangers, Future Anthropologies', *Anthropology Today* 18: 3–5.
Sahlins, M. (1990) *Ilhas de História*, Rio de Janeiro: Jorge Zahar Ed.
Sahlins, M. (2008) "The Stranger-King; Or, the Elementary Forms of Political Life", *Indonesia and the Malay World*, 36(105): 177–199.
Villela, J. (2004) "A Dívida e a Diferença. Reflexões a respeito da reciprocidade", *Revista de Antropologia*, 44(1). São Paulo: USP: 185–220.
Wagner, R. (1981) *The Invention of the Culture* (revised and expanded edition), Chicago: University of Chicago Press.

6 Negotiating access to an Argentinean military institution in democratic times

Difficulties and challenges

Alejandra Navarro

Introduction

Field access and access to interviewees, understood as a dynamic and flexible process, is a fundamental stage in social research (Hammersley and Atkinson 1994; Goetz and LeCompte 1988; Descombe 1998; Feldman *et al.* 2003). Its features, dilemmas and challenges have been the object of methodological debates in different disciplines, such as anthropology, which deploy ethnographic studies. Attention to access has been less frequent in non-ethnographic qualitative studies. However, in this kind of study, access is also of paramount importance in order to establish good relationships with research participants and gatekeepers.

This article analyzes the complex decision-making process that allowed me to interview Argentinean military officers of two military institutions in 2008.[1] The two military institutions are the Military School (Colegio Militar de la Nación) and the Superior Teaching Institute of the Army (Instituto de Enseñanza Superior del Ejército).

The following presentation is divided into three parts. First, I will define "access," drawing upon theoretical and methodological literature. Second, I will briefly describe the complex set of interactions with key informants and potential gatekeepers over almost two years that preceded my final access to the military institutions. The result of this extended negotiation had implications for the development of the study. I will give an account of this process as a way to illustrate the credibility of the research work (Maxwell 1996; Whittemore *et al.* 2001; Cho and Trent 2006). In the last section, I will reflect upon the "false illusion" that access to an institution implies the cooperation of all of its members. The fieldwork experience with a group of military officers confronted me with this reality.

Defining "access"

Within a social research study, access to an institution involves more than obtaining permission to walk in through the front door. According to Glesne and Peshkin, access refers to the process of "obtaining consent to go where you want, watch what you want, talk to whoever you like, obtain and read the materials you want, and do all this for as long as it takes to answer the research questions" (1999: 33). From this perspective, access is only obtained when the researcher can watch and talk freely with anyone in order to arrive at answers to the research questions. According to the British and American methodological literature, it is possible to find at least two tiers to this process. On the one hand, we have to be able to enter and stay in the field (Hammersley and Atkinson 1994), but it is also necessary to establish meaningful relationships with members of the researched institution in order to co-produce a narration (Feldman 2003). According to Wanat (2008), building trusting relationships with and obtaining cooperation from research participants are crucial aspects of access. It is useful to analyze this stage of research work, paying attention to researcher identity, as well as the different presentations of the self it is necessary to put into practice over time (Harrington 2003).

From this perspective, access implies a "formal" entry into the institution, as well as establishing a good "rapport" with research participants to "dig from their memory" (Sautu 1999) those recollections that will provide "clues" in response to the research questions.

It is interesting to analyze different aspects of access. First, authors have examined access as a relational process embedded in institutional meaningful contexts. In other words, access is a task that implies a relationship between research and research participants (Hammersley and Atkinson 1994). Moreover, the researcher needs to take into account the microsocial characteristics of the studied group or institution, including its members, their interactions and the meanings they attach to them. Looking at these particularities is a useful framework, both to unpack the complexities of the negotiating process and to contextualize the analysis of the data obtained.

The second aspect to consider refers to the necessity to plan strategically according to the moments, places and people (Lahmar 2009). During the fieldwork one faces different situations which require analysis of the adequate steps to carry out in order to enter, to remain and to obtain information from the "authorized voices."

Third, various authors point out the importance of the researcher's ability (personal professional abilities) to deal with the ongoing and varied challenges that negotiating access involves (Hammersley and Atkinson 1994; Feldman *et al.* 2003). From this point of view, it is necessary to take into consideration not only the biographic and demographic characteristics of the researcher, but also his/her skills and strategies which are brought into play in order to obtain access. Thinking of us as part of the research process involves using reflexivity as a way to increase awareness of oneself and of others. To do so, it is necessary to take into account what we mentioned in the first dimension, and to be aware of the nature of the institution we wish to enter, and its actors.

Finally, researchers have argued that the ethical dimension of any research process needs to be carefully monitored. The stage of "access" also involves particular ethical challenges and dilemmas. It is necessary to assure confidentiality and anonymity regarding everything that was discussed in the field, as well as the academic treatment of the information. The way in which the study was presented, its institutional framework and the researcher's insertion and affiliation are all key to the negotiation of access into any type of institution, and particularly the military.

The following section examines the process of negotiation with the military institutions that lasted almost two years. I will describe each decision and the steps I had to consider to obtain the final consent to interview officers of the Argentinean Army.

Looking for access and cooperation

The first steps

For this research it was not possible to begin the fieldwork without permission to interview. A military institution is a bureaucratic and hierarchical organization. Taking this into account, I wondered with whom I had to initiate contact, since I did not know anyone in the Army.[2] In order for me to pursue a Ph.D. thesis, the main question initially seemed to concern how to begin. At that moment, my research interest was to analyze the way in which young military officers attributed meaning to their role as military men (their identity). This was valuable in the context of the important educational, normative and organizational reforms that were taking place in the Argentinean Armed Forces. The new missions and the vanishing of conflict hypothesis seemed to draw the military towards tasks closer to those of civilians and not men of arms (Moskos *et al.* 2000). I wanted to deepen the evaluations and meanings attributed by young military officers to this new institutional reality.

Several tasks began simultaneously. First of all, I had to have a clear understanding of the characteristics of the institution I wished to enter in order to plan an entrance strategy. The revision of the existing literature, both theoretical and empirical, heightened my understanding of the particularities of the military organizations; which are, as I have just said, formal and hierarchically stratified institutions. Briefly, we can state that given their size and nature, these are socially complex organizations; they have a public nature and are highly differentiated in the interior. Also, the extensive literature points out that there are few complex organizations with such capacity for control over the individual actions of its members (Abrahamson 1985; Janowitz 1985). Therefore, it is possible to understand the Army as a bureaucracy in which there are no formal counter-power mechanisms and limitations of legal authority.

It was essential to be clear on these particular features in order to define who to approach and identify as the "gatekeeper" to introduce me to the rest of the actors. It did not seem to be intelligent to gain access through the contact of a subaltern or chief officer, notwithstanding the fact that I did not know any. I remembered that in the year 2006, I had been invited to teach a Research Methodology Seminar, alongside other colleagues, at the Superior War College (Escuela Superior de Guerra). I decided to contact the current Academic Secretary of that School and talk to him about my interests. I presented myself and disclosed my institutional affiliation and the aims of my study. That officer agreed to the interview and contacted me with some of his peers. "That was easy!" I thought; I had my research work. I felt that I had begun the fieldwork. I was pleasantly surprised as I thought that entering into a military institution would be much more difficult. My preconceptions relating to the group seemed to be wrong given that with one phone call I had scheduled an interview. Almost immediately, I realized this was not the case. At that first meeting I faced a very kind officer who did not answer my questions because he needed his superior's consent to do so. I faced the characteristics of an institution I thought I knew well. I had an overwhelming sense of disappointment: I had research questions, but nobody would answer them, at least not in their current construction. One interviewee after another explained to me that they could not discuss such topics and all of them suggested that I talk to their superiors.[3] But I did not want the view of a superior officer. "Where did I go wrong? Did I wrongly elaborate my instrument? Were these not the right questions? Did I need to rethink the access strategy?" These were some of the reflections in my field notes. The practical restrictions were modifying the topic of my research and its design.

As the field narrowed due to the impossibility of interviewing other officers and the reluctance to discuss certain subjects, I decided to focus on the revision of the normative and internal documentation. This task could provide me with insight into the changes in the functions and missions of the Armed Forces, while I rethought the strategy for accessing the interviewees. Also, reviewing my field notes in which I registered not only the dynamics of the interview situation, but also my feelings and emotions, I identified my own fear of asking certain questions. I repeatedly wrote phrases such as "I didn't want to ask," "I could not deepen," "I felt that I was being bothersome and did not ask." These notes put me on alert, although at that moment I could not identify their relevance.

During the revision of the normative and the reading of a specific bibliography, I identified a vacancy in military sociological research. Starting from this finding, incorporating the difficulties I found in the field and my own difficulties in relating to the interviewees, I reoriented my original research questions. I was not only interested in the values attributed by Army officers to the introduction of an educational and labor project in the Armed Forces, but also in their sociodemographic features. I started questioning the characteristics of people who wanted to start a military career: where their families were from, what type of studies they had pursued, what were their jobs. I wanted to design a research study focused on the sociology of the military and deepen the biographies of different cohorts of military officers to compare different socio-historical moments. The last study of this type had been conducted in Argentina in the 1960s (De Imaz 1964). This important vacancy deserved further research.

Rethinking the research project and the issue of access

Considering the changes that the military institution was undergoing, and the important gap in the knowledge relating to certain aspects of military sociology, I decided to rethink my research. I now wished to analyze the social bases of recruitment of three cohorts of Argentine Army officers, the social and marital links constructed in their biographies, as well as the meaning attributed by the interviewees to the military career and the choice of a military career.[4]

Once again, I found myself facing the dilemma of who to establish the first contact with and how. In order to answer my new research questions, it was necessary to contact army officers of three cohorts (according to their entry into the Military School) and from different branches of the Army. My previous experience informed my decision to now approach the military in a manner respectful to their hierarchy. In this case, my first contact was a high-ranking official, a former General, who was carrying out functions in the Superior Teaching Institute of the Army (Instituto Superior de Enseñanza del Ejército), an educational institution of the Army. Not only did it seem a good way to begin, but it was also the only contact I had.

Sometimes at the beginning encounters are casual and unplanned. How did I access this retired officer who became one of the "insider informants" (Harrington 2003: 612) of the work? I met him through a researcher of the University of Buenos Aires who knew about my interests and had conducted meetings with this General regarding science and technology. The fact that this officer knows and works in the academic world was highly beneficial; due to this fact, he was possibly more inclined to accept our first meeting and listen to my research proposal. In addition to this, the researcher who mediated the contact is recognized in the academic world and spoke favorably of me. Both situations (knowing the academic world and having someone well known recommending me) were very positive in initiating contact. Almost two years had passed since my first approach to the theme and my first interview.

It was mid June in the year 2008 that the first negotiations for the access into the military world were consolidated; and yet at that moment, I did not know precisely which institutions were going to allow me to enter to do the interviews.

Starting from this first contact with the former General, I began a road of encounters and missed encounters with different officers until I could initiate the fieldwork. All the time, I tried to show myself as concerned in the interests of the institution opening its doors to me.

The long road to the interviews

1 The first informant: former General.
2 The second informant: former Colonel (Secretary of Evaluations of the Superior Teaching Institute of the Army).
3 The third informant: another former Colonel (Member of the Superior Teaching Institute of the Army).
4 The thesis project and my CV had to be evaluated by the Headmaster of the Superior Teaching Institute of the Army.
5 Meeting with the Academic Secretary of the Superior Teaching Institute of the Army (Active Colonel). The fourth informant.
6 Military School: Active Colonel, Secretary of Academic Research.
7 Superior Teaching Institute of the Army: biographical interviews.
8 Superior War School (*Escuela Superior de Guerra*) and the Superior Technical School (*Escuela Superior Técnica*): the survey.

In that first meeting with the former General, I was anxious and nervous. My preconceptions relating to the military were present and indicated that I was going to find obstacles to doing my work. The surprise was absolute and immediate. I faced a man dressed as a civilian, very kind and easy-going, completely interested in academic life and the integration of civilians and the military. Before anything else, I introduced myself and made my institutional affiliation explicit, which became an important method of access. Right away, we started discussing my research objectives and the importance of these objectives for the military institution. Analyzing the social bases of recruitment of the Army in the last 50 years seemed attractive and did not seem as likely to generate rejection. I had to accrue the information which supported this fact. I explained to the General that apart from revising the statistics, I wanted to perform biographical interviews with officers of different cohorts to deepen knowledge of family context and lifestyle, as well as the motivations for choosing a military career. At this point, the most complex issue was explaining the importance of the strategy and approach, and I lost space for negotiation in the interview situation. In fact, as I registered in my field notes:

> In a certain point of the conversation I found myself guided by the interests of my "sponsor." The first feeling I had was that I was going to end up doing what he/she wanted. I was upset; I did not know what to do, or how to explain why it was so important to investigate those

> issues. I listened attentively to the explanations given to me and immediately explained again what I wanted to do and why it was necessary to do biographic interviews and not a survey. I felt I handled the situation again. Now I have to build a schedule.
>
> (Field diary, 23/06/08)

It is interesting to think about the benefits and dangers of "gatekeepers." They are an important gateway, but can also make us refocus the research to where they want it to go (Goetz and LeCompte 1988; Hammersley and Atkinson 1994; Feldman *et al.* 2003). As Harrington expresses (2003: 598) "the power of the participants in defining the themes that are important to the researcher" turned out to be an important exercise in repositioning myself as a researcher and making the argument about the importance of the study's objectives. In this task, my identity and the identity of the informants were present (Harrington 2003). I was a woman, relatively young, in a masculine and hierarchical world, conversing with a General of the Nation who cordially indicated to me what was most interesting to the institution: "At times I feel I cannot point out anything out of place and I have to tell them what they want to hear. I am very aware of my words almost all the time. It is interesting watching myself vacillating" (Field diary, 23/06/08).

Finally the meeting ended, on good terms; it seemed that I was now able to start my fieldwork. The General contacted me with another former officer, a Colonel who was working as the Secretary of Evaluations of the Superior Teaching Institute of the Army. In this new meeting, I explained once again what I needed from them: access to statistics and contact with several officers to interview. I told the Colonel I wanted to reward them in some way for all the help they were giving me. I offered to teach a Research Methodology Course at the institute, or to give a conference of their choosing.[5] This officer pointed out that it was not necessary for me to teach a course, that they were interested in the conducting of research concerning military institutions in order to achieve a higher integration between the civil and the military worlds.

For approximately two months, several encounters with both informants took place, and a third informant appeared: another Former Colonel member of the Superior Teaching Institute of the Army. In each meeting, I once again took account of my research objectives and repeated my lack of understanding regarding why I could not start the fieldwork. My field notes were fundamental. These allowed me to reflect and understand the importance of the characteristics of the institution, as well as the assumption that one should have a "portfolio" of identities to put into motion each time one establishes contact with one of the informants. As these encounters took place, one of the informants managed my access to the institution. The thesis project and my curriculum vitae had to be evaluated by the provost of the institution.

For a long time I had a feeling of uncertainty. It seemed I would be able to start the interviews but only with my established informants, and it was not clear how I was going to contact others. At the same time, it seemed that the statistical records with the information about the bases of recruitment would be impossible to access. This finally happened, and I had to change part of my objectives. As time went by, the dynamic character of the negotiation became clearer, as did the importance of internalizing the microcosm of the institution. I took care not to neglect the hierarchical line and the formal aspects of the organization were always present. I was very careful with my appearance and I paid attention to my clothes. I always dressed formally and never wore trousers.

While I was waiting for formal access to some institution depending on the Superior Teaching Institute of the Army, the three officers I had spoken to previously allowed me to interview them. I viewed this as a good opportunity to test the interview guide. For approximately three months, I introduced my work and myself. By mid-September (after the first meeting in June) I was told I had to make a formal presentation of my work to the Academic Secretary of the Institute. This active officer was not only very kind, but he also showed interest in the project and the theme. The conversation was long and I had the opportunity to explain each of my research objectives, giving account of the approach each would require. I felt very comfortable and I assumed my role as a researcher defending the relevance of my research purposes, without losing sight of the flexible and sensitive interests of the group I wished to enter. The officer offered me the opportunity to conduct a survey in two institutions: the Superior War School and the Superior Technical School. This was an opportunity to access another military organization.[6]

This long meeting ended on excellent terms. The officer also committed to gaining permission to start the fieldwork in the Military School (the only Officer School in Argentina). I became aware that I had started my negotiation to access the interviewees and that they were going to be officers working in some military institute depending on the Superior Teaching Institute of the Army. The characteristics of my informants (their work places) would shape my cases (those who I would interview). I understood the importance this had, as well as the importance of the particularities of these centers. In my work I would be giving accounts of the outlook of that particular group and not of the officers of the Army as a whole. In my field notes, I found myself reflecting on the selection of the cases and their characteristics:

> The Academic Secretary talks to me about the possibility of interviewing at the Military School. I thought I would be able to interview officers in different barracks. It appears that will not happen. I have to think about how to justify that choice and its importance and pertinence. I am not clear on that yet. I will see what happens.
>
> (Field diary, 08/25/2008)

The next step was going to the Military School. I was nervous and anxious. I had an interview with an active Colonel, who was Secretary of Academic Research. He was expecting me with the endorsement of the Academic Secretary of the Institution of Higher Education of the Army, which the Military School depended on. I felt as though I was starting all over again. The officer, however, was very nice and was interested in the subject of my research.[7]

This visit was useful in many ways. The opportunity of meeting with the officer and the visit to the educational establishment allowed me to learn a little more about how the Military School functioned. On the other hand, I could specify a schedule of interviews based on the officers' assignments there. It was also possible to confirm that the statistical data I needed to study the social bases of recruitment in the last 50 years was not there. I did not have the statistical data, but I was on my way to starting the fieldwork and the biographical interviews of the officers of the Military School.

At the end of October, I conducted the first biographical interview, starting from the agenda I had made with the Secretary of Academic Research. The literature points out that to "achieve access perseverance and flexibility are important" (Feldman *et al.* 2003: 27). Alongside the fieldwork it was necessary to accommodate the decisions of my "gatekeeper" in the Military School. He was the one indicating whom to interview and I did not have any leeway to select the officers. My field diary gives an account of my own reluctance to suggest interviewing a particular officer and accepting every decision made by the Colonel.

The way in which the field developed had clear implications for the research process. For example, I had not originally thought of interviewing women, but I did. I also only wanted to interview officers of the army and I ended up interviewing officers from the other services. Not having enough time to transcribe and perform a previous analysis dictated that I be much more detailed in the constructions of the interview memos and field notes which replaced that initial preliminary analysis so common in qualitative research.

While I was performing the biographical interviews in the Military School, I realized that it would be difficult to interview senior officers there (officers who entered the institution before 1973). It was necessary to access another institution. I went back to my fourth informant in the Superior Teaching Institute of the Army and managed to do some interviews there as well. The population was completely different; almost all were former officers, but still pertinent to my research objectives. The methodology was similar. Each time I left, the informant (Secretary of Academic Research) managed to organize the agenda and put me in touch with the officers. Occasionally, I had to reprogram the interviews because of lack of time, since the interviewees extended the conversation. The difference I felt was that these former officers handled themselves with more freedom in relation to the informant, and despite "being military men" were more distended regarding certain formalities (expressing personal opinions, verbal attributions, extending the conversation). At times I felt I was backstage in the military world, whereas in the Military School I was more in the "front stage" (Goffman 1994).

The biographical interviews were being done. I had managed to access a world perceived as distant and different from my own.

Access is not always to cooperate: "the false illusion"

Despite that feeling, a last fieldwork experience towards the end of the year 2008, the application of a self-administered ***survey*** to a group of chief and subaltern officers (students in the Superior War College and the Technical College), confronted me with the false illusion that to access is always to co-operate.

The fourth informant of the Superior Teaching Institute of the Army put me in touch with an active Colonel, the Academic Secretary of the Superior War College. He was in charge of gathering all the officer students to hand out the survey and wait for them to complete it.[8] When I arrived, I presented my institutional affiliation, the aims of the project and myself. The situation was tense. I made it very clear that the survey was voluntary and anonymous and that the treatment of the information would be strictly academic. Nobody refused to answer it, but showed opposition when they left several questions blank. I felt very uncomfortable. The situation became even worse with the arrival of the Secretary of Evaluation. This officer had not been informed of my visit and was very annoyed and aggressive, overruling the quality of my work. Once the survey was finished, I could not ask respondents to deepen their responses. I considered that for an exploratory phase, and given the lack of studies of that type, it was a good start. After that experience, I decided not to repeat the survey with the other cohort.

This experience illuminated the real dimension of access, which is much more complex than the achievement of a good rapport with the informants. Also, my "self" and personal characteristics (abilities or inabilities) to negotiate with the actors were present, as were the power relations to which, at times, I felt at a disadvantage.

Final thoughts

The experience was enriching and allowed me to understand the military institution and its members. The process of identification, initiation and maintaining of the contacts has to be documented as completely as possible. This experience gives data, tells us about the characteristics of the institution, object of the study and contextualizes the scenario for the interpretations. It is interesting to keep the relationships established between the members of the institution and the researcher present. The tenacity and the willingness to accept some of the lines of action of the informants (for example, the selection of cases) were key. For more than a year I worked on this access and I talked to different hierarchies. The clear explication of my objectives helped the process and so did the professional and academic insertion. Although some studies point out that in the initial stages the researcher is in a position of power (Karmieli-Miller *et al.* 2009), in my case, the situations varied and at times the informants positioned themselves with more strength, changing the direction of the study's focus.

I could overthrow the myth that it is not possible to conduct a study within the military institution if you are not a native. I had to work on my own preconceptions, but this happens with all research topics. Negotiating access is simply building relationships with different actors and each interview situation is particular and unique. The respect for the voice and outlook of the other, beyond our own valuations, is key to positioning ourselves as researchers who are plural, flexible and interested in giving an account of the inter-subjectivity and multiplicity of the social world.

Notes

1 This chapter is part of my Ph.D. research, "A Look into the Biographic Trajectory of a Group of Officers in Argentina: Class Origin, Marital Bonds and Motivation in the Choice of a Military Career", University of Buenos Aires.
2 In my previous research work in the same area, I interviewed Army officers who had been discharged and convicted (Navarro 2007, 2009). My contact with them had been casually established.
3 At this early stage, I held three interviews and some informal conversations.
4 The cohorts respond to the year of leaving the Military School (Colegio Militar de la Nación). The first cohort is made up of officers who graduated from Military School before 1973 (exercised their careers when the Armed Forces had a leading role in the country); the second is made up of officers who graduated between 1974 and 1986 (served as officers during the last military government in Argentina); and the third are officers who graduated during the democracy.
5 Some authors (Blau 1964 and Wax 1952 in Harrington 2003) give accounts of the exchange approach in access negotiations. It focuses on the compensation given to participants in fieldwork.
6 This was a sub-group to which I was only going to apply the self-administered survey, looking into similar aspects of the biographical interview, but without the same level of depth. This way I got information from a larger group and analyzed the value they attributed to their ingress into the Superior School of War and the Superior Technical School.
7 Before starting the conversation, and respecting the hierarchical line, I introduced myself to the director of the Military School who opened the doors of the institution and suggested that it would be useful for them to know the results of my study.
8 To get this meeting I had to make several calls (at least three) until we arranged a day.

References

Abrahamson, B. (1985) "La profesión militar y el poder político: los recursos y su Movilización," in R. Bañòn and J. Olmeda (eds.) *La institución militar en el Estado contemporáneo*, Madrid: Alianza Universidad, 254–269.

Cho, J. and Trent, A. (2006) "Validity in Qualitative Research Revisited," *Qualitative Research*, 6(3): 319–340.

De Imaz, J.L. (1964) *Los que mandan*, Buenos Aires: Universidad de Buenos Aires.

Descombe, M. (1998) *The Good Research Guide for Small-scale Social Research Projects*, Berkshire: Open University Press.

Feldman, M., Bell, J. and Berger, M. (2003) *Gaining Access: A Practical and Theoretical Guide for Qualitative Researchers*, Walnut Creek, CA: AltaMira Press.

Glesne, C. and Peshkin, A. (1999) *Becoming a Qualitative Researcher*, Reading, MA: Addison-Wesley.

Goetz, J.P. and LeCompte, M.D. (1988) *Etnografía y diseño cualitativo en investigación educativa*, Madrid: Morata.

Goffman, E. (1994) *La presentación de la persona en la vida cotidiana*, Buenos Aires: Amorrortu.

Hammersley, M. and Atkinson, P. (1994) *Etnografía. Métodos de Investigación*, Buenos Aires: Paidós.

Harrington, B. (2003) "The Social Psychology of Access in Ethnographic Research," *Journal of Contemporary Ethnography*, 32(5): 592–625.

Janovitz. M. (1985) "La organización interna de la institución militar," in R. Bañòn and J. Olmeda (eds.) *La institución militar en el Estado contemporáneo*, Madrid: Alianza Universidad, 101–139.

Karmieli-Miller, O., Strier, R. and Pessach, L. (2009) "Power Relations in Qualitative Research," *Qualitative Health Research*, 19(2): 279–289.

Lahmar, F. (2009) "Negotiating Access to Muslim Schools: A Muslim Female Researcher's Account on Experience," paper presented to the Postgraduate Research Student Conference in Nottingham on 14 July 2009.

Maxwell, J.A. (1996) *Qualitative Research Design: An Interactive Approach*, Thousand Oaks, CA: Sage.

Moskos, C., Williams, J. and Segal, D. (2000) *The Postmodern Military: Armed Forces After the Cold War*, Oxford: Oxford University Press.

Navarro, A. (2007) "Matrices y Tipologías en el análisis cualitativo de datos: una investigación con relatos de oficiales Carapintadas," in R. Sautu (ed.). *Práctica de la investigación social cuantitativa y cualitativa*, Buenos Aires: Lumiere, 301–323.

Navarro, A. (2009) "Looking for a New Identity in the Argentinean Army: The Image of the 'Good Soldier'," in C. Dandeker, G. Caforio and G. Kuemmel (eds.) *The Military, Society and Politics: Essays in Honor of Juergen Kuhlmann*, Schriftenreihe des Sozialwissenschaftlichen Instituts der Bundeswehr, Netherlands: VS Verlag, 59–74.

Sautu, R. (1999) *El método Biográfico*, Buenos Aires: Lumiere.

Wanat, C. (2008) "Getting Past the Gatekeepers: Differences Between Access and Cooperation in Public School Research," *Field Methods*, 20(2): 191–208.

Whitemore, R., Chase, S. and Lynn Mandle, C. (2001) "Pearls, Pith and Provocation: Validity in Qualitative Research," *Qualitative Health Research*, 11(4): 522–537.

7 Research relations in military settings

How does gender matter?

Helena Carreiras and Ana Alexandre

Introduction

The question of how gender impacts the research process has frequently been addressed in the methodological literature of the social sciences. Much less has been written on the way that it affects the development of research in a particularly gendered organization such as the armed forces.

The aim of this chapter is to discuss the way gender impacts research in military settings, highlighting both commonalties and differences regarding other fields of study and the variety of aspects through which gender might affect the course of the research. A number of issues are examined, resorting to concrete research examples: the impact of the researcher's and the researched gender on negotiating access or discursive interaction during interviews; the gendered interpretations of the researched; the gendered nature of the context and the gender focus of the research topic.

After revisiting proposals that refer to the research process as a social process itself, the chapter proceeds with a revision of the literature on the "gender factor" in research and gendered nature of military settings. It then moves onto the examination of a concrete research experience – a case study of a Portuguese peacekeeping battalion deployed to a NATO-led mission in Kosovo in 2009 – in order to illustrate how, in this specific case, researchers dealt with the gender dimension of the research process, including both the trade-offs involved in the negotiation of their roles and forms of control required to acknowledge and minimize the impact of gender in the conduct of field research.

Research relations as social relations

Challenging an epistemological paradigm which assumes that the so-called "researcher bias" can be neutralized by adhering to a traditional positivist model of sociological research, qualitative researchers have underlined the socially interactive nature of the research process and the need to reflect upon the role of the researcher in its course. A long list of accounts by researchers from different social scientific disciplines has shown that the research process is a much more complex enterprise than just the result of a linear application of methodological receipts or technical options. As Burgess puts it, "research is not just a question of neat procedures but a social process whereby interaction between researcher and researched will directly influence the course which a research programme takes" (Burgess 1984: 31).

Even if this observation applies to different types of methodological strategies, including quantitative survey research, it becomes particularly clear in the practice of qualitative research and in the conduct of field research. Becker (1996) noted that there are two circumstances that are likely to produce perceived differences between qualitative and quantitative approaches: on the one hand, the two sorts of methods raise somewhat different questions. While quantitative researchers are interested in the establishment of relations between variables, qualitative researchers aim at unfolding a system of relationships, looking at the way how things hang together in a web of mutual interdependence. On the other hand, the environment where they work is usually very different. Since field research predominantly involves the use of observation, unstructured interviews and documentary evidence that must be applied to a specific social setting, qualitative researchers cannot insulate themselves from the data or avoid establishing relationships with other actors in the setting. Most situations in which qualitative researchers find themselves are socially significant events, having major consequences for the research process. The flexibility of procedures, the lack of fixed roles, the absence of a set of rigid rules, which are characteristic of fieldwork, additionally stress the impact of research relations on the research process.

What is being argued here is that any research process involves social interaction, even if in different degrees, depending on the chosen methodological strategy; that, as in any other social interaction, it takes place in social contexts and produces social effects; that finally, reflexivity is essential to acknowledge and control possible bias. Far from being a constraint, the awareness of such effects is the very condition for the production of social scientific knowledge; to ignore them would be to prevent the possibility of asserting the validity and reliability of the research results.

Within this framework, qualitative researchers have been encouraged to reflect on their role in the research process. One central issue in their work, and also particularly relevant for this chapter, has been the way in which the status characteristics of the researcher affect the research process, namely in terms of gaining and negotiating access to the field, and of establishing and maintaining rapport with respondents and informants in a setting. Aspects such as age, sex, ethnicity or socioeconomic status have been particularly highlighted, but other characteristics such as the experience of the researcher, the expectations and assumptions made about him/her by the researched and the power differential between them have equally been stressed (Scott 1984). Referring to this last issue in his reflection about the interview process, Bourdieu (1993) called attention to the negative effects of power asymmetries between interviewer and interviewed. In his view, differences in status – usually to the advantage of the interviewer – might generate situations of symbolic violence that seriously jeopardize the research effort. Therefore, he sustains that social proximity and familiarity are the two main conditions for a nonviolent communication (Bourdieu 1993).

Carrying further this concern, other authors stressed the opposite situation, where power differential seems to be at the disadvantage of the researcher (McKee and O'Brien 1983; Lee 1997). In both cases, and as a consequence, the advice to seek for symmetry in interview situations has been vehemently put forward. For instance, drawing upon her experiences interviewing women about motherhood, Oakley (1981) made the proposal that interview relationships should be nonhierarchical and that the researcher should be prepared to invest her personal identity in the relationship. However, this particular piece of advice has also raised skeptical reactions. Other authors called attention to the contradictions that emerge if the principle of symmetry is blindly accepted and noted that perfect congruity is not only rarely possible in interviewing, but may not even be desirable. The same characteristics that in some cases facilitate access might, in other instances, be detrimental to the research effort:

> Certain status characteristics of the researcher may facilitate gaining a access to a setting in virtue of their non threatening nature.... [However,] the same characteristics which work to the researcher's advantage in terms of gaining access may become liabilities when the focus shifts to establishing and maintaining rapport with respondents.
> (Gurney 1985: 58)

This begs the question of which of the many social characteristics at issue are the most important to a particular situation (Riessman 1987: 191) and underlines the fact that there are benefits to being both an "insider" or "outsider" to an experience by virtue of one's group membership (Merton 1972).

The issue of multiple positionality of the researcher and the question of how his or her different social attributes mediate research relationships thus adds further complexity to the discussion. How have these questions been discussed with regard to the gender factor in research?

The gender factor in research

Among the growing body of literature that explores the social nature and impact of research relations, a significant number of studies address the gender dimension of the research process. Issues such as the way gender influences access to the field, negotiation of the researcher's roles, establishing rapport with respondents, the questions that are posed and the data that is collected in field projects, have been examined in detail (Warren and Rasmussen 1977; Wax 1979; Golde 1986; Whitehead and Conaway 1986; Warren 1988).

Since the early 1970s, many of these discussions have focused on the issue of the *researcher's gender*, and particularly upon the question of how being female influences the roles to which the researcher is allocated and how this may limit or impede the progress of research (Easterday *et al.* 1977; Burgess 1984: 90–91).

For ethnographers, especially, the question has often been that of distinguishing between what might be problems, regardless of gender, common to fieldworkers from difficulties unique to women (Golde 1986; Bell *et al.* 1993). It has been found that women researchers are often allocated roles which are consistent with the stereotypical picture of women, a pattern that puts them in a subordinate position that might, in turn, inhibit their access to a variety of situations. However, existing empirical accounts have not been short of paradoxes and often highlight contradictory dynamics. For instance,

> A female researcher ... may discover that if she is to gain access to privileged information, she must play a subordinate role which requires some deception and improvisation.... For another colleague, such deception may represent an unethical and unviable alternative, yet, either course of action may significantly alter what is studied.
>
> (Hondagneu-Sotelo 1988)

Likewise, assessments of gender in interview situations have provided interesting and contradictory evidence regarding the particular effects of single or mixed-gender interviews. Many of those who use "same-gender" interviews base this preference on the intuitive notion that rapport is more easily achieved in these contexts. Drawing upon her own experience of interviewing women on motherhood, Finch (1984) underlined the distinctive character of the woman-to-woman interview, describing how her female informants showed a high degree of trust and expected her to understand them because of the shared gender of interviewer and interviewee.

Other researchers have challenged this assertion. Analyzing two contrasting interviews with an Anglo and a Puerto Rican woman, conducted by a woman interviewer, Riessman (1987) concluded that gender congruence did not help the middle-class white, Anglo interviewer make sense of the working-class Hispanic woman's account of her marital separation. In this case, gender congruity was not enough to overcome ethnic incongruity. The potential bonds generated by the fact that both interviewer and interviewee were women were insufficient to create shared meanings that could transcend divisions between them: "The lack of shared norms about how a narrative should be organized, coupled with unfamiliar cultural themes in the content of the narrative itself, created barriers to understanding" (Riessman 1987: 173). This led the author to conclude that beyond the researcher's gender, conditions for sensitive collaboration should be evaluated.

Two other studies, aimed at evaluating the impact of the interviewer's gender on the data generated, bring additional elements to this reflection.

Williams and Heikes (1993) examined men's responses to a male and female interviewer using data from two independent in-depth interview studies of male nurses. The authors discuss the possible impact of the researcher's gender in more general contexts than the same-sex situation and suggest that the gender of the interviewer is not an insurmountable barrier to establishing rapport and achieving reliable results in in-depth interviewing. They note that

> the "definitions of the situation" conveyed by the men in our two studies showed remarkable similarity and overlap, even on topics involving gender and sexuality, which have been identified by survey researchers as the topics most sensitive to "sex of interviewer effects."
>
> (Williams and Heikes 1993: 289)

Likewise, in a study of the work and family experiences and aspirations of young adult women conducted in 1992, Padfield and Procter draw a systematic comparison between interviews conducted by a man and a woman, and although there was a marked difference in the voluntary addition of further personal experience, they found a consistency between the two interviewers in responses to questions asked about the sensitive topic of abortion (Padfield and Procter 1996).

What these studies suggest is that putting the emphasis exclusively on the gender of the researcher oversimplifies the debate, while at the same time underline the diversity of aspects in which gender might become relevant in research.

One of these aspects regards the *gendered roles and interpretations of the researched,* since perceptions of subjects also define the researcher's role, and here gender plays a vital part. Padfield and Procter note that

> the interviewee is an active participant in the definition of the interview process. Interviewers may take account of gender in the conduct of interviews. The same applies to interviewees who also define the situation in gender terms. The problem is to know just how the interviewees' sense of gender is influencing the interview.
>
> (Padfield and Procter 1996: 364)

By taking into account the respondents' negotiation of the gendered context of the interaction, it is possible to challenge the assertion of woman-to-woman identification, as well as proscriptions against cross-gender research.

Additionally, gender is but one of the attributes that has an impact on research relations. Multiple social characteristics intersect with gender and mediate research relationships. In her work interviewing male corporate elites, Schoenberger (1991) argued that a shared "class" status marked her recognition as "one of them" despite her position as a woman (1991: 281). Likewise, in an interview-based study of a male elite population at London City banks, McDowell (1998) described how, besides gender, age was a significant attribute in her positioning towards the interviewee. McDowell highlighted the fact that her multiple positionality as a white, heterosexual, middle-aged woman interviewing a range of younger people, mainly white, and for the most part explicitly heterosexual, was equally intervening in the negotiation of the relationship.

Beyond the gendered identities of the researcher and the researched, the gendered perceptions of the researched or the intersection of gender with other social attributes, the *topic of the research* has also been considered a crucial factor in shaping research relations. In the literature, this has been mainly addressed with regard to topics that might be defined as "sensitive," in the terms proposed by Renzetti and Lee (1993), that is, topics that potentially pose for those involved a substantial threat, the emergence of which renders problematic for the researcher and/or the researched the collection, holding and/or dissemination of research data (Renzetti and Lee 1993: 5). For instance, drawing on her experiences of interviewing men on workplace harassment, Lee (1997) addresses the topic of interviewer vulnerability when discussing sexualized topics in the context of gendered interview dynamics. In response to what is identified as the underplaying of this concern in existing discussions of interviewing, the author details her own repertoire of personal safety strategies and analyzes the dilemmas regarding control, rapport and reciprocity that arose from these tactics (Lee 1997).

This is why Schwalbe and Wolkomir (2001: 91) suggested that to examine the impact of gender on interviews we need to move beyond "Who is asking whom?" to "Who is asking whom about what?". Pini (2005) carries this critique further to reflect upon the *gender focus of the research* and also upon the importance of *the research environment*. In an interview-based study of rural and agricultural workers and masculinity in an Australian sugar industry company, she argues that, in order to examine the way in which gender may shape an interview, we need to go beyond a simple focus on the gender of the researcher and the researched, and undertake a more sophisticated analysis which explores the intersection of the mediating influences of "who, whom, what and where."

Drawing on the work of Schwalbe and Wolkomir (2001), Pini argues that beyond the focus on the "who," "whom," and "what," the "where" should be of a central concern because the gendered context of the research environment also informs the interview relationship:

> In using the term "where" I do not simply mean the room, building or organization in which an interview takes place, although the physical site of the interview is another factor mediating the enactment of power relations (Elwood and Martin 2000). The "where" to which I refer is the broader field or context in which the research is taking place.
>
> (Pini 2005: 204)

If this broader context of research refers to sex-typed occupations or male-dominated settings, the issue of context becomes extremely relevant. One common conclusion of a variety of studies of women researchers in such environments is that when conducting research in male-dominated places, such as a police force or the military, the gender status of the researcher assumes greater significance.

Building upon her own study of a prosecutor's office, Gurney (1985) underlined the fact that "some researchers may never succeed in achieving more than superficial acceptance from their respondents because of the status each researcher occupies." She points out that "some female researchers studying male dominated groups frequently find themselves in just such a position or do not attempt to gain entry in certain male dominated settings. Sex roles expectations may hamper women's work in the field" (Gurney 1985: 42).

Processes of encapsulation, which ensue from stereotyping, have been found to characterize the experience of minority women (or other minorities) in organizations (Kanter 1977), especially if the respective occupations are gender-typed and the presence of tokens is felt as intrusive (Blalock 1970; Yoder 1991). If research relations reproduce or emulate general patterns of social relations, similar processes might also develop with relation to women researchers who work in gender-typed contexts.

After noting that most instructional literature on qualitative research suggests that the novice fieldworkers should start by adopting a passive, nonthreatening, incompetent role vis-à-vis setting members and then turn to a competent and knowledgeable professional role once accepted by the group, Gurney gives a good example of how different it might be for a woman or a man to achieve this in a male-dominated context:

> while stereotypical attitudes towards females generally assure their acceptance in the naïve incompetent role ... those same attitudes hamper females' efforts to make the transition to the professional role. Female researchers must work especially hard to achieve an

> impression combining the attribute of being nonthreatening with that of being a credible, competent professional.
>
> (Gurney 1985: 43)

The same pattern has been underlined in research on the police:

> It may be that the female researcher entering this type of environment will face similar dilemmas to the policewoman. Unlike the policewoman, the researcher does not have necessarily to prove herself capable of facing rioters, or breaking up a fight, but she does need to prove herself competent and, above all, trustworthy.
>
> (Horn 1997: 299)

Particularly concerned about the extent to which a woman researcher might experience sexist remarks, sexist behavior and sexual hustling, in male-dominated settings, Gurney proposed that: (a) the female researcher should try to project a professional attitude through not only outer appearance, but demonstration of research skills and her ability to convey knowledge concerning her chosen field of specialization; (b) that special attention should be devoted to the way how women are treated in the setting, since the probability exists that women researchers will be subject to the same type of behavior; (c) to take advantage of the marginality status of "not being one of the guys" which may enhance her awareness of prejudice and discrimination in the setting (Gurney 1985: 59).

Exactly in the same vein, and apparently profiting from Gurney's advice, Horn found in her research that it was possible to use the perceptions of the researched in her advantage, which, however, raised ethical questions and influenced the data collected.

Another aspect where there might be significant gender differences and which has to do with the specificity of military settings, regards the level of secrecy usually associated with multiple instances and practices within these organizations. It has been claimed that institutions such as the police or the military have high levels of secrecy and suspicion might arise that the researcher will be a spy. According to Hunt (1984), since women tend to be associated with the clean management world, the assignment of the researcher to the role of spy is more likely if the researcher is a woman.

Is it thus important to ask: To what extent is gender a salient component in military contexts and why?

The military as a gendered organization

One reason why gender is a salient factor in military settings is because the military, like other formal organizations, has a particularly clear gender regime, even if it is less homogeneous than is usually supposed. The military can be considered a gendered organization, in terms of the criteria that scholars have used to classify organizations as gendered (Britton 2000). First, the military's organizational structure is clearly based on gender divisions, both in terms of opportunity and power (hierarchical divisions) and in terms of occupational structure (sexual division of labor). Women are excluded from certain functions and there are distinct patterns of gender representation by rank and functional areas. Second, it is male-dominated in terms of numeric representation, especially in the areas more closely related to the core functions of the institution, exactly those that confer not only more prestige and rewards, but also objective possibilities to access the higher hierarchical ranks. Despite a notable increase in the representation of women throughout the organizational structure over the last few decades, it is a realistic assumption that male dominance will continue to exist in the near future. Finally, from the point of view of culture, hegemonic definitions of military conflate with hegemonic masculine culture and ideology, even if such construction is subject to historical change and varies significantly in different sectors inside the institution. In any case, the military has for centuries been a source of normative conceptions of gender, which, on the one hand amplifies dominant cultural patterns and, on the other, actively participates in its production and reproduction. Thus, more than merely gendered, the military has also been seen as a "gendering," gender-granting, or gender-defining institution (Cohn 1993; Segal 1999); it amplifies, as in a magnifying glass, the social dynamics of gender (Reynaud 1988).

However, there is a notable heterogeneity of gender environments in the military. Some units and branches are more gender integrated than others and this division maps onto a hierarchy of masculinities. Cohn provides a suggestive example of such diversity. In her study of gender and national security, she found that while combat unit men tended to think of themselves as the most "studly," officers in combat support and combat service-support offered a different understanding, suggesting that the macho masculinity was really a compensation for combat soldiers' lack of technological and organizational skills. Consequently, she adds, "in different branch and unit contexts, men construct different components of the construct 'masculinity' with concomitant differences in their attitudes about women in the military" (Cohn 1999: 35). This is exactly why the military has to be understood in terms of the relationship between masculinities (Connell 1987, 1995). It is the relationship between forms of masculinity – some physically violent but subordinate to orders, others dominating and organizationally competent – that may help us understand the present state of gender relations in military organizations. In fact, it would be inaccurate and naive to suppose that military operations actually work on the basis of traditional heroism archetypes. With the technologization of warfare, the "management of legitimate violence" became strongly based on rational-bureaucratic techniques of organization. Due to the multiplication of support functions, in modern armies the majority of soldiers are not combatants at all and most military leaders would agree that "Rambo types should not be driving our jeeps and supply trucks" (Connell 1995).

Additionally, the presence of women has challenged the gendered characteristics of this environment. From the mid 1970s, the historical pattern of women's exclusion from regular military participation has been dramatically challenged. Contrasting with the exceptionality of their prior involvement in warfare, women have started to be admitted in the armed forces in peacetime and with full military status. By the beginning of the twenty-first century, all NATO countries had admitted and increased the number of women in their armed forces. Although restrictions existed, many were lifted; women were progressively allowed to enter military academies and given access to a wider variety of positions and functions, gender awareness grew within most military structures, and integration policies were designed and implemented (Carreiras 2006, 2010).

Independently of how extensive we consider the impact of this new pattern of women's military participation to be, one thing seems clear: this is no longer a change only for the duration of conflicts, as previously happened. However unequal their status or occupationally segregated and culturally discriminated, women are no longer peripheral to the armed forces. To some extent, the amplification of women's military roles has meant a challenge to the common view of the armed forces as a male domain and the male-warrior paradigm (Dunivin 1994).

Various types of missions can also differently build on gender archetypes, depending on a variety of factors, such as the mission's specific goals and rules of engagement, the way that pre-deployment training has raised (or failed to raise) gender awareness among soldiers, characteristics of the local contexts and patterns of civil–military interaction in the field.

Gender and field observation: a case study of a peacekeeping mission in Kosovo

The research chosen to illustrate some of the gender dynamics discussed before is a case study of a Portuguese infantry battalion deployed to a peacekeeping mission in Kosovo in the framework of the NATO-led Kosovo Force (KFOR) between March and September 2009. This case study was developed in the framework of a larger research project on the Portuguese Armed Forces after the Cold War, aimed at understanding the organization's structural and cultural adjustment to change from the end of the 1990s.

Drawing on Pini's proposal, we will focus on the "who," "whom," "what," and "where" questions, to analyze this research experience and explore some aspects of how gender mattered and how it interacted with other factors in the conduct of research.

It is useful to start with the "how" question, that is, the specific research design chosen for this case study. Awareness of the potential impact of gender on the research project should lead to conscious options during the preparatory phase of any research, namely the definition of the research design and its operationalization.

How? The research design and the preparation of fieldwork

The specific research design for this case study consisted of a mixed-method strategy, involving a variety of methodological procedures and tools. Between February and October 2009, a research team composed of a senior female researcher and two junior male and female researchers, followed a Portuguese infantry battalion, from the pre-deployment phase in the infantry regiment in Vila Real, a city located in Northern Portugal, to their return home after accomplishing a six-month mission as Kosovo Force Tactical Reserve Maneuver Battalion (KTM) in Kosovo.

During the pre-deployment phase, we ran a survey on all 292 members of the battalion, had formal and informal meetings with commanders and high-ranking officers, conducted semi-structured interviews with men and women of different ranks, seniority and occupational areas, as well as with a selected sample of soldiers' wives.

During the deployment phase, the research team spent two weeks with the battalion in a military camp in Pristina, Kosovo, participating as much as possible in its daily activities. This included sharing meals and sleeping quarters, attending field exercises, watching the soldiers' regular activities around the camp, going to the gym, attending the bar and Saturday night karaoke sessions, participating in routine events and ceremonies as well as in special occasions at the KFOR level, etc. We also interviewed them individually and in a group, in formal and informal ways.

Finally, the post-deployment period consisted of a shorter stay at the batallion's headquarters in Vila Real, back in Portugal, a couple of weeks after its return from the mission. Then, again, an evaluation survey was run on the whole battalion and more informal meetings and exchanges took place.

Various decisions that were taken during the preparation of the research already incorporated a gender perspective. While designing the research, we acknowledged the way gender would matter, especially in the phases of research that required more direct and extended contact with the soldiers. This was clear in the way decisions were taken regarding "who interviews whom," attributing interviews with staff officers to the senior researcher (her previous experience had shown that at this level, and probably due to a more symmetric relationship in terms of formal qualifications and power, gender effects are attenuated) while having the younger researchers interviewing enlisted personnel, privileging, whenever possible, same-sex interviews.

Knowing that the period of field observation would be much more demanding in terms of involvement and intensity of relationships, the decision was taken to keep, train and deploy a gender-mixed research team. As shall be addressed further on, this was considered essential to: dilute the potential encapsulation that could occur if the team was exclusively composed of female researchers (a majority of the research team); facilitate the integration of the group; reduce concerns and suspicion from the researched; and allow for a broader observation platform.

Another example refers to the way, while preparing access to the field, we acknowledged the results of previous research, as well as the answers to the pre-deployment survey about gender relations in the military and, particularly, those concerning the presence of women in international missions.

If it is true that there are no recipes, that it is virtually impossible to fully anticipate the way how a field mission will evolve and previously design the research in an ideal manner, existing knowledge and accounts of other research experiences were of great help here. For instance, we tried to negotiate beforehand the sleeping arrangements during the field mission so that the researchers would not be accommodated far from the battalion. Military contexts are extremely structured and it is not easy – although not impossible – to make changes to previously defined arrangements and rules. This is even more so in special situations such as operational field missions like the one that we were about to experience. Our presence was inevitably a source of tension for the commanding officer who later on confessed that the idea of dealing with the mission requirements and rules of engagement, while at the same time dealing with a team of "strangers" whose security had to be guaranteed (and whose goals were still not totally clear) was, in his words, a "headache."

Therefore, at the level of the preparatory phases of the research, strategies to deal with the impact of gender on research in military settings include: previous awareness, impact evaluation and adjusting research options more explicitly than in other contexts.

What? The topic of the research

The research was designed to explore a variety of analytical dimensions related to the overall research framework. The issue of women in the military or their participation in the mission was but one among a variety of research topics, namely the internationalization of the Portuguese Armed Forces, soldiers' experience of international peacekeeping missions, evaluation of the present mission (including aspects such as organizational issues, daily activities, rank relations, sociability, relationship with other forces and local populations, communication with the family), as well as attitudes towards the military and its transformation, professional identities and expectations and conciliation between family life and the military profession.

The gender focus of the research was thus mitigated. Specific research questions in this analytical dimension included the following: How do men and women describe and evaluate the performance of gender-mixed units in international missions? What attitudes and understandings do they develop about peacekeeping and international missions? How do they manage conciliation of their professional and family lives? To what extent do these new contexts of operation affect perceptions of women's military roles?

In sum, the topic of the research was not particularly sensitive from a gender perspective, even if we felt that, as in many other research experiences, gender would necessarily have implications both for what was disclosed or withheld, pursued or neglected during the interviews, informal gatherings and conversations with the soldiers.

Who? The social characteristics of the researchers (and the gendered expectations about them)

The research team was composed of three researchers of different ages and seniority: one senior researcher in her forties, the coordinator of the study; a young female research assistant and an equally young male researcher, both in their late twenties. As noted before, the choice of this composition was not arbitrary: to have a mixed-sex group, including a young male researcher, was crucial for the image we wanted the battalion to have of the "sociologists" who were coming to share their field experience in Kosovo and avoid the encapsulation effects that could arise if there were only female researchers. And indeed, the fact that we were given different dormitory spaces in the camp – a direct "gender effect" – was extremely important to amplify the visibility zones and observation platforms over the daily activities of the battalion.

"Gender of the researcher" effects were generated in a variety of circumstances.

As in other research experiences in military settings, we felt that it was easier to have men talk about emotions and intimate issues when the interviewer was a woman. For instance, many of them seemed to enjoy the possibility of expressing their concerns about family and marital relationship at a distance, with all the tensions and anxiety this provoked; and they did it more extensively with the female researchers than with the male researcher.

Another example regards the particular position of the senior female researcher. Given her previous work on the integration of women in the armed forces, there was an implicit expectation that she would be a supporter of gender integration: women used her to voice discontent, sometimes overemphasizing problems or difficulties of the integration process; but at least in some cases, this expectation also helped to build complicity. Men, on their part, tended to avoid negative references to women, but this did not prevent many of them showing a critical attitude over the issue of women in combat or the effects of romantic relationships on mission performance. There was, however, a constant justification of such viewpoints. Interestingly though, this track record in the study of gender in the military – a supposedly "soft" topic – did not serve to raise the senior researcher's status as an expert in studying the military. Her credibility was often under test. However, the expectation that she would not know things about the military universe was probably less related to gender than one might have expected, since the young male researcher also experienced the same attitude; it was probably more the result of another structural operative characteristic in this context: the fact that we were civilians and none of us had served in the military. As Castro notes in his fieldwork among cadets in a military academy (addressed in his chapter in this volume), being a civilian is a military invention, a differential characteristic of the qualitative researcher in a military institution.

But besides this major status difference between the researchers and the researched, our multidimensional positionality (age, sex, authority) helped to build relationships and generate bonds of various kinds. Access and rapport were built at different paces, rhythms and moments; with some groups it hardly happened, with other it was more extensive. Nevertheless, different roles and complicities were developed on the basis of those varying status characteristics.

Where? Physical and social spaces and the gendered culture of the research context

It has been argued that peacekeeping missions are contexts where the military has been subject to a process of "degendering," that is, redefinitions of the soldier's role have taken place to incorporate an additional humanitarian dimension on the basis of which new relational and communicational skills and capacities are valued in a way that overcomes traditional gender archetypes usually more evident in extreme operational contexts. For instance, studies conducted in Southern Europe show that increased participation in peacekeeping missions has led to a redefinition of the contents of traditional professional roles among certain sectors of military personnel (Batistelli 1997; Carreiras 1999, 2011).

However, research results have also been contradictory. While a majority of studies on gender and peacekeeping stress the relationship between successful mission performance and the increased involvement of women (Karamé 2001; Hendricks and Hutton 2008; Bridge and Horsfall 2009), others still underline the cultural contradictions surrounding the construction of the peacekeeper's role and the tensions that ensue from that process with possible negative effects on performance (Sion 2008).

In the case of the battalion we followed, women represented 11.4 percent of the force, a percentage that can be considered high if compared to the usually lower female representation in military operations.[1] This numerically significant presence of women, together with the fact that this was a low intensity mission, may account for the rather relaxed atmosphere we experienced from the perspective of gender relations. But this did not mean that gender was not being permanently "done." In one of the most challenging situations for the researchers, the soldiers tested us on explicit gender grounds. During our first Saturday karaoke session at the camp, "sociologists" were called to the stage to sing a very profane "sexualized" popular song. The dilemma was thus the following: on the one hand, we knew that this was a decisive test to gain access and trust, especially among enlisted personnel; failing it would probably have negative consequences for our integration and acceptance; on the other hand, there was an obvious discomfort with the situation, especially on the part of the female researchers. We overcame the dilemma by resorting to a silent dancing performance while the young male researcher sang the song. The test was passed successfully and we were praised for our "courage."

As far as physical space was concerned, in the context of operational field missions, it is not uncommon to find disruptions to existing segregation patterns, such as having more common spaces without rank restrictions. For instance, we found out that there were shared sanitary premises in the women's dormitory area, extremely rare in military units back in Portugal. However, hierarchy, one of the central pillars of the military organization, was still clearly displayed. We were required to eat together with the officers and noncommissioned officers who used a different refectory. Although at the beginning we felt that it would have been interesting to enjoy more freedom to share meal times with the enlisted, this was obviously a nonnegotiable issue. We were under the control and supervision of the commanding structure; to question this arrangement would have been felt with suspicion by the officers but probably also as intrusive by the troops.

The sleeping quarters for women were a much freer space. Because of logistical constraints, all women were accommodated together (officers, NCOs and enlisted). This created situations where there could be more direct contact. One day, we sat on the wide corridor floor and had a rather open group interview, avoiding the more formal space given to us for running the semi-structured interviews. This proximity, which was only possible because of our gender status, allowed us to overcome the salience of rank divisions, which would have been felt in a much stronger way, had the spatial arrangements been different.

On the other hand, however, in terms of movements around the camp, there was a clearer concern towards the two female researchers than the male researcher, who could use more freedom of movement. As other researchers noted, a woman might be seen as less threatening but this may also put her in the role of requiring more help and surveillance. But here again, we all felt that problems of access or control derived not only from our gender status but clearly from the interplay between this positionality and our status as "civilian."

Conclusion

Because military organizations have been defined as gendered organizations, research in military contexts is supposed to be affected by gender in particular ways. Research experiences in male-dominated occupations have reported the amplified impact of this variable on the conduct of research. In this chapter, we reviewed a variety of aspects in which gender might become relevant in research, and provided illustrative empirical evidence on its specific impact on research in a military setting, drawing on the sociological study of a peacekeeping mission in Kosovo.

We sustain that, while it is impossible to avoid "gender effects" in research relations in military settings, the extent to which such relations become more or less gendered depends on the interplay of a variety of factors that are simultaneously cognitive and social. These include the following: (a) methodological strategy and research design; (b) gender of the researcher and the researched and the way this interacts with other status characteristics: age, perceived seniority and authority; ethnicity, etc.; (c) gender understandings of the researched and the gendered context of the research environment; and (d) the gender focus of the research. It is the interplay of these factors and not any one of them alone that generates concrete patterns of influence of gender on research.

Therefore, to identify and control gender effects requires a context-sensitive analysis that takes into account the various dimensions in which gender impacts on the research process. As Hondagneu-Sotelo puts it, "By acknowledging choices and the social context in which they were made, the researcher can identify bias, monitor its effects on the data, and develop a sharper analysis" (Hondagneu-Sotelo 1988).

One way to do this is to carefully exercise sociological reflexivity throughout the development of research and from its preparatory phases.

Testing "if" clauses might be a promising avenue for this endeavor. For instance, it is possible to put forward the hypothesis that if the research topic is not particularly gender sensitive, if the context is not strongly gendered, if the research objectives do not require full immersion methodological strategies, and if conditions are created for sensitive communication and trust between the researcher and the researched, it is likely that gender will not have a strong impact. Contrarily, if all the opposite clauses hold, it is probable that gender will have a very profound impact on research. In between situations will emerge from a varying composition of these factors. The challenge for researchers is thus the explicit incorporation of gender impact evaluation and reflexivity throughout the research process.

Note

1 Levels of female participation in military missions vary significantly, but in general the proportion of women soldiers in the military component of peacekeeping missions is still much lower than their representation in the respective national armed forces. Data from the UN Department of Peacekeeping Operations (UNDPKO) shows that female presence ranges from a mere 1 to 4 percent of the military personnel in these missions (UNDPKO 2010) while the percentage of women in military forces goes up to 20 per cent (Carreiras 2010).

References

Babbie, E. (2007) *The Practice of Social Research*, 11th edition, Belmont, CA: Thomson Higher Education.

Battistelli, F. (1997) "Peacekeeping and the Postmodern Soldier," *Armed Forces and Society* 23(3): 467–484.

Becker, H. (1996) "The Epistemology of Qualitative Research," in R. Jessor, A. Colby and R. Shweder (eds.), *Ethnography and Human Development: Context and Meaning in Social Inquiry*, Chicago: University of Chicago Press, pp. 53–71.

Bell, D., Caplan, P. and Karim, W. (eds.) (1993) *Gendered Fields – Women, Men & Ethnography*, London: Routledge.

Blalock, H. (1970) *Towards a Theory of Minority Group Relations*, New York: Capricorn.

Bridge, D. and Horsfall, D. (2009) "Increasing Operational Effectiveness in UN Peacekeeping: Toward a Gender-Balanced Force," *Armed Forces & Society* 36(1): 121.

Bourdieu, Pierre (1993) *La Misére du Monde*, Paris: Seuil.

Britton, D. (2000), "The Epistemology of Gendered Organizations," *Gender & Society* 14(3): 418–434.

Burgess, R. (1984) *In the Field – An Introduction to Field Research*, London: Unwin Hyman.

Carreiras, H. (1999) "O que Pensam os Militares Portugueses do Peacekeeping?" *Estratégia* 14: 65–95.

Carreiras, H. (2006) *Gender and the Military: Women in the Armed Forces of Western Democracies*, London: Routledge.

Carreiras, H. (2010) "Women in the Armed Forces of Western Democracies," in J. Buckley and G. Kassimeris (eds.), *Ashgate Research Companion to Modern Warfare*, London: Ashgate.

Carreiras, H. (2011) "Gendered Culture in Peacekeeping Operations," *International Peacekeeping* 17(4): 471–485.

Cohn, C. (1993) "Wars, Wimps, and Women: Talking Gender and Thinking War," in M. Cooke and A. Woollacott (eds.) *Gendering War Talk*, Princeton, NJ: Princeton University Press, 227–246.

Cohn, C. (1999) *Wars, Wimps and Women: Gender in the Construction of US National Security*, Berkeley, CA, manuscript.

Connell, R. (1987) *Gender and Power: Society, the Person and Sexual Politics*, Stanford: Stanford University Press.

Connell, R. (1995) *Masculinities*, Berkeley, CA: University of California Press.

Deutsch, F. M. (2007) "Undoing Gender," *Gender & Society* 21(1): 106–127.

Easterday, L., Papadermas, D., Schorr, L. and Valentine, C. (1977) "The Making of a Female Researcher: Role Problems in Field Work," *Urban Life* 6: 333–348.

Elwood, S. and Martin, D. (2000) "Placing Interviews: Location and Scales of Power in Qualitative Research," *Professional Geographer* 52(4): 649–657.
Finch, J. (1984) "It's Great to Have Someone to Talk to: The Ethics and Politics of Interviewing Women," in C. Bell and H. Roberts (eds.) *Social Researching: Politics, Problems, Practice*, London: Routledge.
Golde, P. (ed.) (1986) *Women in the Field – Anthropological Experiences*, Berkeley, CA: University of California Press.
Gurney, J. (1985) "Not One of the Guys: The Female Research in a Male-dominated Setting," *Qualitative Sociology* 8(1): 42–62.
Hendricks, C. and Hutton, L. (2008) "Defence Reform and Gender," in M. Bastick and K. Valasek (eds.) *Gender and Security Sector Reform Toolkit*, Geneva: DCAF, OSCE/ODIHR, UN-INSTRAW.
Hondagneu-Sotelo, P. (1988) "Gender and Fieldwork," *Women's Studies Int. Forum* 11(6): 611–618.
Horn, R. (1997) "Not One of the Boys: Women Researching the Police," *Journal of Gender Studies* 6(3): 297–308.
Hunt, J. (1984) "The Development of Rapport Through the Negotiation of Gender in Field Work Among Police," *Human Organizations* 43: 283–296.
Jessor, R., Colby, A. and Shweder, R. (eds.) (1996) *Ethnography and Human Development: Context and Meaning in Social Inquiry*, Chicago: University of Chicago Press.
Kanter, R. (1977) *Men and Women of the Corporation*, New York: Basic Books, Inc.
Karamé, K. (2001) "Military Women in Peace Operations: Experiences of the Norwegian Battalion in UNIFIL 1978–98," in L. Olsson and T. L. Tryggestad (eds.) *Women and International Peacekeeping*, London: Frank Cass, 85–96.
Lee, D. (1997) "Interviewing Men: Vulnerabilities and Dilemmas," *Women's Studies International Forum* 20(4): 553–564.
McDowell, L. (1998) "Elites in the City of London: Some Methodological Considerations," *Environment and Planning A* 30: 2133–2146.
McKee, L. and O'Brien, M. (1983) "Interviewing Men: Taking Gender Seriously," in E. Gamarnikow, D. Morgan, J. Purvis and D. Taylorson (eds.) *The Public and the Private*, London: Heinemann, 147–161.
Merton, R. (1972) "Insiders and Outsiders: A Chapter in the Sociology of Knowledge," *American Journal of Sociology* 78: 9–47.
Oakley, A. (1981) "Interviewing Women: A Contradiction in Terms," in H. Roberts (ed.) *Doing Feminist Research*, London: Routledge, 30–61.
Padfield, M. and Proctor, I. (1996) "The Effect of Interviewer's Gender on the Interviewing Process: A Comparative Enquiry," *Sociology* 30: 355–366.
Pini, B. (2005) "Interviewing Men: Gender and the Collection and Interpretation of Qualitative Data," *Journal of Sociology*, 41(2): 201–216.
Renzetti, C. and Lee, R. (eds.) (1993) *Researching Sensitive Topics*, London: Sage.
Reynaud, E. (1988) *Les Femmes, la Violence et l'Armée*, Paris: Fondation pour les Études de Defense Nationale.
Ridgeway, C. and Correll, S. (2004) "Unpacking the Gender System: A Theoretical Perspective on Gender Beliefs and Social Relations," *Gender and Society* 18(4): 510–531.
Riessman, C. (1987) "When Gender Is Not enough: Women Interviewing Women," *Gender and Society* 1(2): 172–207.
Riessman, C. (1991) "When Gender Is Not Enough: Women Interviewing Women,"

in J. Lorber and S. Farrell (eds.) *The Social Construction of Gender*, London: Sage, 217–236.

Scott, S. (1984) "The Personable and the Powerful: Gender and Status Is Sociological Research," in C. Bell and H. Roberts (eds.) *Social Researching: Politics, Problems, Practice*, London: Routledge & Kegan Paul.

Schoenberger, E. (1991) "The Corporate Interview as a Research Method in Economic Geography," *Professional Geographer* 43(2): 180–189.

Schwalbe, M. and Wolkomir, M. (2001) "The Masculine Self as Problem and Resource in Interview Studies of Men," *Men and Masculinities* 4(1): 90–103.

Segal, M. (1999) "Gender and the Military," in J. Chafetz, *Handbook of the Sociology of Gender*, New York: Kluwer Academic/Plenum Publishers, 563–581.

Sion, L. (2008) "Peacekeeping and the Gender Regime: Dutch Female Peacekeepers in Bosnia and Kosovo," *Journal of Contemporary Ethnography* 37(5): 561–585.

UN Department of Peacekeeping Operations (UNDPKO) (2010) *Gender Statistics as of February 2010*. Online. Available at: www.un.org/en/peacekeeping/contributors/gender/2009gender/feb09.pdf.

Warren, C. (1988) *Gender Issues in Field Research*, Newbury Park, CA: Sage Publications.

Warren, C. and Rasmussen, P. (1977) "Sex and Gender in Field Research," *Urban Life* 6(3): 349–369.

Wax, R. (1979) "Gender and Age in Fieldwork and Fieldwork Education: No Good Thing Is Done by Any Man Alone," *Social Problems* 26: 509–522.

Whitehead, T. and Conaway, M. (1986) *Self, Sex and Gender in Cross-cultural Fieldwork*, Urbana: University of Illinois Press.

Williams, C. and Heikes, J. (1993) "The Importance of Researchers' Gender in the In- depth Interview: Evidence from Two Case Studies of Male Nurses," *Gender and Society* 7(2): 280–291.

Yoder, J. (1991) "Rethinking Tokenism: Looking Beyond Numbers," *Gender and Society* 5(2): 178–192.

8 Inside the military organization

Experience of researching the Slovenian Armed Forces

Janja Vuga and Jelena Juvan

Introduction

The methodology today known as 'qualitative' has been used for centuries, although by different names (e.g., fieldwork, ethnography, etc.). Participant observation has been used over the centuries by anthropologists and sociologists who believed that the best way to gain insight into people's attitudes, relations, etc. is to live with and observe them. Researching a military organization has proven to be extremely challenging. For example, it is very difficult to gain permission to enter such an organization and even harder to be given access to some specific units (special forces, military intelligence). Even after a researcher enters the military, they are confronted with yet another obstacle, namely, gaining the trust of the servicemen/servicewomen. The researcher's presence itself presents another problem. It is almost impossible to ensure the 'normal' functioning of the military when there is an 'intruder' in its midst. This can be partially overcome if the researcher is a member of the military organization. However, this could also raise the issue of trust since other servicemen/women might have strong reservations about discussing their problems with someone who is part of the same institution due to the fear of such information being passed on to their superiors. This might be a problem, in particular, if the researcher has a higher rank. The question of the impartiality of a researcher who is also a member of a studied population also arises. A similar situation can be identified in the Slovenian Armed Forces (SAF) where servicemen/women feel very reserved about consulting a military psychologist in case they need help since every consultation is marked in their personal files and, thus, might influence their future military career. The presence of civilian and neutral researchers coming from an outside research institution might solve some issues, yet at the same time it might provoke some other issues and dilemmas of researching military organizations. Some of the problems encountered when researching the SAF are described and further explored in this chapter.

Qualitative analysis 'in theory'

Qualitative research differs from the quantitative variety. While the aim of quantitative analysis is to gain exact, reliable and objective cognitions, qualitative analysis discovers new relations; the research object enters the researcher's consciousness with his/her own perspective and thus provokes the researcher's self-reflection (Huzjan 2004: 187). While quantitative analysis aims to verify or disprove existing theories, qualitative analysis often derives theory from data gathered in the field (e.g., grounded theory). Creswell (1998: 15) defines a qualitative analysis as a process during which the researcher gains the whole picture by analyzing stories and own observations, while trying to lead the research in a natural environment. However, the first problem occurs when defining a so-called natural environment in the research context, practically an impossible task since every research process produces unnatural situations during its course. This means that the researcher, along with the notion of being the object of the research, produces situations different from usual ('normal') (Kogovšek 1998: 24). This does not necessarily mean that the research is not valid or objective (Alasuutari 1995: 94). To accept this, it is necessary to overcome the methodological approaches used in quantitative analysis.

Despite its long tradition and frequent use in several disciplines (e.g., sociology, anthropology, etc.) several theories still argue about the validity and reliability of data gathered by qualitative research methods.[1] Qualitative research is characterized by its holistic nature since the object of observation is not observed partially but as a whole (Taylor and Bogdan 1984). It should be naturalistic, meaning that the researcher should not disturb the natural process of the observed group. However, the latter is more of an ideal than a reality since people always change their behavior when they are being observed or studied (Kogovšek 1998: 52). Further, qualitative research is inductive and researchers are discovering new theories and concepts based on their experiences in the field. Naturally, the researcher is aware that his or her own cultural background, prejudices, beliefs, religion, etc. will influence the understanding of the observed group and the relations between them. When employing qualitative analysis, one should also be aware that the research cannot be repeated (Kogovšek 1998: 54) because social reality changes as we speak. Findings gathered by fieldwork are always subjected to what the object of the researcher's observation is willing to share. Moreover, Neuman (1994: 357) states that the ability to repeat research is not proof of validity since repeatability is impossible when observing social interactions or society as a whole. Alasuutari (1995: 92) connects the validity of the researcher's interpretation of a certain situation with his or her ability to logically explain why people react in a certain way.

One can identify several factors influencing the validity of data acquired by qualitative methodology. One of them is the relationship between the researcher and the object of research (Kogovšek 1998: 63). In the case of observation with participation, the presence of a researcher will clearly influence the behavior of individuals, as well as the interactions among members of the observed group. There are two types of observation, namely 'with' or 'without' participation.[2] The method of observation is appropriate for observing a specific group of people, their habits, relations, attitudes, the interactions among them, etc. Yet researchers should take into account that the findings cannot be applied to other groups nor can they be generalized. Observing with participation enables researchers to use different methods, both quantitative as well as qualitative, which can lead to a better understanding of a certain phenomenon.

Ways to understand the 'hearts and minds' of servicemen and servicewomen

Studying military organizations shows that the use of more than one method is crucial due to one of its specifics, which is a closed environment. It is very difficult to gain permission to enter a military organization and even more difficult to encourage servicemen and servicewomen to cooperate and share their problems. It is impossible to say whether their answers are honest and therefore observation may be considered a useful tool to estimate the reliability of acquired data. Besides that, the researcher can never move freely within military barracks or gain access to all places. It should also be taken into account that the 'story' presented to the researcher is often well-prepared. However, observing life within military walls, even with some evident restrictions, helps researchers create a clearer image and improve their understanding of the information gathered with other methods. Participation with observation can also help develop the framework for interviews (Dean *et al.* 1969).

One method that can be, and in our case often is, successfully merged with observation with participation is an interview as a face-to-face method where the first impression that is gained is particularly important as the time the interviewee and researcher spend together is very limited. A structured[3] interview is a method known in both qualitative and quantitative methodology and used very often in researching military organizations, especially when there is a sample of a few hundred respondents. It can also be used to compare opinions held by several individuals on a specific topic (Vogrinc 2008: 107). Data gathered by this method can be easily coded and interpreted with the use of predefined categories (Fontana and Frey 1994). On the other side, when we want to gain individuals' opinions, attitudes, perspectives, etc. we use a nonstructured interview. The researcher roughly prepares an idea of what needs to be discussed, with no specific questions. There is still a rule that the researcher should not offer own opinions and should therefore stay neutral and out of the discussion (Kogovšek 1998: 33). However, Fontana and Frey (1994) believe just the opposite, namely that the researcher should participate actively in the interview to guarantee reliability and honesty. In between, there is the semi-structured interview, which basically means that a researcher has a list of topics but the exact wording of questions and their sequence is formed during the interview (Kogovšek 1998: 31). Further, Sagadin (1995: 318) and Tashakkori and Taddlie (1998: 102) developed a special type of interview that merges some characteristics of the structured and nonstructured interview that is the so-called funnel strategy (in Vogrinc 2008). In line with the latter, the researcher proceeds from general questions at the beginning to more complex and closed ones towards the end of the interview. The researcher thereby creates a relaxed environment and gives the interviewee time to prepare for more difficult questions. Patton (1990) defines yet another type of interview that stands between the structured and semi-structured interview, that is, the so-called standardized open-ended interview. The same questions are posed to all interviewees in the same order. Those questions are open, which represents the main distinction between a standardized open-ended interview and a standardized interview. To pretest a questionnaire or to conduct an in-depth analysis in the field a group interview can be used. Its specific feature is a wide selection of opinions that lead the researcher to clearer conclusions and a better understanding of certain phenomena (Fontana and Frey 1994). The researcher must be experienced and know how to engage every member of the group, including those who are less talkative. Group interviews are an important source of information; they are flexible, cumulative and relatively cheap. However, the researcher needs to prevent situations whereby some interviewees step out and impose their opinions on others. A problem can also occur when some members of the group do not feel comfortable sharing their opinions in the presence of others (ibid.). The latter happens relatively often in instances when group interviews are being conducted with members of a military organization.

After years of analyzing military organizations, based on experiences with the SAF it can be established that the standardized open-ended interview combined with a structured interview and observation with participation yield the optimal insights and enable an in-depth analysis of a military organization. In this manner, the problem of ordered participation in the research is minimized.

Despite all efforts, an interview is not a neutral tool for acquiring data. The answers are influenced by the context; in the social sciences, the researcher's personal attributes are very important (e.g., gender, ethnic background, social class, etc.) (Kogovšek 1998: 29). In addition, Kogovšek (1998: 34) states that social stratification can be reflected in an interview. Denzin (in Kogovšek 1998: 34) claims that an interview is influenced by established gendered identities and information is therefore limited by those same identities. When talking about a military organization, gender is particularly important. However, this depends on the ratio of men and women entering the armed forces in general, the relations between the two genders, as well as the gender ratio within a studied unit. In most armed forces, combat units are formed only by men and therefore those units evolve in a masculine environment. It is general knowledge that a military organization is highly masculine, meaning that in most cases men feel superior to women. When there is a female interviewer and a male interviewee, a shift in roles occurs (ibid.). Still, the presence of a female researcher could be disrespected by the objects of research or they might even not take the research seriously, making their answers irrelevant. Conversely, some members of combat units could, due to their highly masculine orientation, accept a female researcher better than a male one and feel more relaxed in her presence. They might also see their mother, spouse or sister in her and therefore develop a positive attitude.

Further, in military organizations, the place where the interview takes place is highly important. Without doubt, an interview organized in the relaxed environment of a faculty would provide different results compared to one organized in military barracks. In every situation, members of the military organization represent their institutions and therefore their answers are limited by certain rules and restrictions. However, conducting an interview in barracks might result in an even more reserved attitude. In the case of military organizations it is very important to create an environment that inspires trust and where the interviewee will feel comfortable discussing relevant topics. This requires a well-trained and experienced researcher. Members of a military organization can only participate in research based on a direct order of their superior, meaning that it can never be clear whether they were willing to cooperate or did so merely as a result of the order. This dilemma is particularly salient since the interviewee's perception of his or her own role in the research significantly influences their answers and consequently the results of the analysis. When conducting an interview with a serviceman, the researcher also needs to take account of the fact that the object of observation will probably not tell the complete truth and uncover personal thoughts or beliefs connected to the studied topic. Members of armed forces speak in the name of their organization and, even in their free time, they represent their military organizations. Military socialization is one of the strongest forms of socialization within a certain organization and being a member of armed forces occupies a high position in soldiers' identities.

In researching the SAF, the gender of the researcher has not been presumed to be an obstacle and female researchers have not reported any kind of disrespect from the male servicemen included in the research. On the contrary, female researchers have found their gender to be a benefit since servicemen were more willing to 'open their hearts' to female researchers and to offer more insight into their personal and family problems.

Impact of the triangulation of methods and researchers

Triangulation is employed to create a better understanding of certain data or merely to see the situation from various perspectives. In the social sciences there is no assurance that research will provide us with reliable and valid data (Kogovšek 1998: 69). Fraenkel and Wallen (2006: 462) claim that a great share of results depends on the expectations, ideas, doubts, prejudices, etc. with which the researchers enter the process. Further, Kogovšek (1998: 69) claims that the use of multiple methods reveals different dimensions of social reality, leading to the conclusion that the application of various methods results in a higher probability of the researcher's correct interpretation of observed phenomena. According to Denzin (in Kogovšek 1998), this can be called the triangulation of methods, researchers, theories and data sources. In the context of researching the SAF, the triangulation of methods and researchers will be emphasized.

Denzin (in Kogovšek 1998) defines two types of triangulated methods, that is, inter and intra. For the analysis of data gathered with quantitative analysis we usually use the intra-method of triangulation.[4] On the other hand we have used inter-method triangulation, which anticipates the use of more than one method when analyzing the same sample, for example for researching participation of the SAF in peace operations. However, a researcher needs to keep in mind that society is fluid and constantly changing and therefore different methods cannot automatically yield identical results even when observing the same phenomena (Lincoln and Guba 1985). Moreover, we try to acquire various interpretations of a situation by using different researchers. Their perception of the environment depends on personal experience and is based on their subjectivity. Therefore, each one might perceive the situation differently or put an emphasis on different gestures, symbols or reactions. Lincoln and Guba (1985: 307) claimed that the main weakness of this method lies in the different perspectives of the researchers. Those authors claimed that different researchers cannot possibly interpret a certain situation in the same way. Accordingly, the findings of one researcher cannot confirm the findings of another. However, the positive aspects of the triangulation of researchers prevail over the negative ones if the limitations are kept in mind.[5]

Besides triangulation, Fraenkel and Wallen (2006: 462) suggest the cross-checking of gathered statements and descriptions. To check the consistency of an individual's opinion, researchers can use multiple interviews of the same person. To avoid misinterpretations and ensure a higher level of validity and reliability, some additional measures are suggested: understanding and speaking the language of the observed group; using a tape recorder or even a video camera; taking notes of the researcher's thoughts and observations during the process; asking additional questions; having the final report reviewed by the studied individuals/groups or a neutral individual; recording exact information about the circumstances in which the interview took place (ibid.).

The role of the researcher

The role of the researcher is much more important in qualitative analysis than in quantitative analysis. With the former, the researcher is part of the studied environment, even though he or she can be more or less actively included. Kogovšek (1998: 55) stresses it is important to decide to what degree the researcher will reveal the dimensions of the study, what amount of time he or she will spend in the studied environment, how actively will the researcher participate, etc. In any case, the researcher's presence will have a certain impact on people's behavior. The researcher also needs to pass by 'gatekeepers'[6] (ibid.). In the case of military organizations, they are usually officers or even higher ranked officers. Their intention, as a rule, is to make the organization 'look good.' They might, therefore, directly influence other members of armed forces or indirectly show them what kind of behavior is expected. Accordingly, officers play a double role; on one hand, they order their subordinates to participate in the research and, on the other, they give them open or concealed limitations. It can be expected that such situations have a certain impact on the dynamics of the observed group and the behavior of individuals. In any case, the most important factor for obtaining high quality information is the attitude the studied group has towards the researcher (Kogovšek 1998: 56).

Qualitative analysis 'in the field'

The Defense Research Center of the Faculty of Social Sciences at the University of Ljubljana has been researching the SAF since 1991, when the Slovenian Armed Forces were established. However, its experience in researching military organizations generally dates further back in the past. As a result, several useful experiences were gained, based on success and also on failure, which helped to evolve defense science. One of the first major projects the Defense Research Center was involved in was a project entitled *The Preparedness of the Political System for People's Defense and Social Self-Protection*[7] which ran from 1983 to 1989. The results of this project are no longer applicable since the concept of people's defense and social self-protection is outdated, yet the experience gained in researching the defense system and military organization during this project has been invaluable and useful for the development of defense science in Slovenia as a whole. In 1990, a project called *Members of the Territorial Defense and Military Profession* was carried out. It applied the approach of Charles Moskos to institutional and professional social organizations in the military for the first time in Slovenia (Jelušič and Grizold 2008: 197). The second project dating from this time was based on the US empirical research *The American Soldier 1941–1945*. The project was entitled *Slovenian Soldier: Members of the Territorial Defense in Direct Combat Contact with the YPA*.[8] Both projects were based on the use of a quantitative methodology. Their main contribution lies in the application of foreign concepts and theories to the Slovenian environment. In 1992, a five-year project called National Security of Slovenia started and included two large empirical studies: *Present and Future of the Military Profession in Slovenia* (1993–1994) and *National Security and International Relations* (1994–1995), which later developed into a regular form of measuring public opinion regarding issues of national security (Jelušič and Grizold 2008: 199). Longitudinal survey Slovenian Public Opinion has given us an opportunity to observe attitudes towards security issues, including the changing cultural patterns and actual matters. In 1995, another project called *Civil–Military Relations in Contemporary Societies* started. After 2004, the number of research projects forming part of the national research program 'Knowledge for science and peace' grew. They were financed by the Ministry of Defense of the Republic of Slovenia and the Slovenian Research Agency. They included several ongoing projects dealing with the SAF. Several projects[9] can be identified as adding to the experience with qualitative research of the SAF.

Experience gained in the long-term research project from 2003 until 2008 dealing with the participation of the SAF in peace operations was very useful for the debate on the appropriateness of applying qualitative methods when researching military organizations. Not only was the research itself based on the use of several different methods, but the scientific basis of the research was also built upon the insights gained from research, based on several different research methods used by the well-known American sociologist Charles C. Moskos among American soldiers participating in Task Force Falcon in Macedonia and Kosovo in September 2000. The study employed various methods: in-field observation, interviews and a standardized questionnaire survey among 320 soldiers. The aforementioned long-term research of the SAF in peace operations was composed of three main research projects.[10] As an important factor in the further use of qualitative methods, the inclusion of SAF members appointed to international headquarters in the research was identified. When researching SAF members appointed to several international headquarters around the world, the standardized open-ended interview was applied. The main methodological concept prepared for implementing the research of SAF in peace operations was in fact a combination of several research methods, both quantitative and qualitative. All the surveys provided measurements at three points in time: before, during and after the assignment. Three time measurements have been very useful for observing the change in motivation, behavior, relations between soldiers, etc. which couldn't be identified by conducting only one measurement. Prior to 2007, when for the first time in its history the SAF participated in peace operations with a whole battalion, measurements before and during the assignment were conducted using a semi-standardized (self-administered) questionnaire, while the last measurement after the assignment was conducted via a face-to-face, semi-structured interview. After 2007, the use of semi-standardized questionnaires was also introduced after the assignment. Measurements during the assignment were conducted in the theaters of operation. During these visits by the research teams, participant observation was also conducted as one of the methods for collecting data and crucial information. In the mission areas, face-to-face, semi-structured interviews were also carried out, mainly with commanders and individual SAF members recruited to the mission's headquarters.

The second very important project which has contributed significantly to the debate on the use of qualitative methods in researching military organizations was the project *Human Resources in the Military* conducted between 2004 and 2006. The project sought to identify the value system of young generations in Slovenia regarding an occupation in the military. It analyzed the values and job satisfaction among those people who had devoted some of their spare time to military training (volunteers for compulsory service and reserve members). It also focused on the problem of maintaining soldiers in the SAF (Jelušič and Papler 2006: 9). The project included eight research groups and 30 experts, with 3,100 civilians and 2,200 service members being the subject of the research. A combination of different research methods was employed in order to achieve the most accurate results possible.

At the Defense Research Center, the triangulation of methods within various researches aimed at observing the SAF has been used in several research projects. Usually, structured and standardized open-ended interviews have been employed in combination with observation with or without participation, and also in combination with group interviews. Merging those methods gives us the opportunity to understand certain data gathered in the standardized questionnaire, also considering the personal experience of the group which was the subject of the research. For example, the majority of results with regard to the SAF's participation show a high level of dissatisfaction with the quality of food during the deployment. For the research team, it was relatively difficult to comprehend this issue when only taking the results gathered from the standardized questionnaires into account. During personal interviews with some SAF members this issue was further elaborated, and more information was given so that the research team could understand why the issue of food during deployment has proved to be the most problematic issue for almost every SAF unit deployed to a mission abroad. Additional information on the situation was given to the researchers during their visits to the area of the mission, explaining that the food dissatisfaction was really a manifestation of range of other problems servicemen and servicewomen were dealing with.

Main questions when researching a military organization

Based on the experiences in researching SAF mentioned above several questions can be raised. Some of them are rather general and can be a characteristic of any other military organization, while others are very SAF-specific.

First, there is the question of entering a military organization as a civilian research institution. By definition and tradition, military organizations are very closed. Rules for entering a military organization are strict and prescribed in detail. A military organization, which as a traditional organization is not very keen on any changes coming from the outside world, is built and organized in such a manner to minimize any influences coming from its environment. The Slovenian Armed Forces do not have their own internal research institution so doing research within the SAF can only be done by an outside civilian institution, which has to be invited (and chosen) via a public tender. These types of projects are financed by the Ministry of Defense and, in some cases, also by a Slovenian research agency. So, an interest for the research has to come from the Ministry of Defense and SAF. It can be mostly noted that the MOD and SAF are prepared to finance only topics and issues which they may find interesting, and not those seen and identified as important by an external research organization. For example, the issues of military families are identified as important among the researches and also among the servicemen and servicewomen themselves; however, no research interest coming from the top of the SAF could have been noticed in previous years, and therefore no financial sources were assigned for this purpose. On the other hand, different approaches are also allowed, although much depends on personal ties[11] with the head of the SAF who is responsible for approving all research projects. And, in those cases, financing the research is a burden of the research institution.

Second, there is the question of establishing trust among SAF members as an outside research institution. Servicemen and servicewomen have proven to be a suspicious and mistrusting population and are very careful about their statements, especially with regard to their relations with higher ranking officers, possible criticism aimed at their superiors and, especially, their personal problems. One can detect a strong fear of the military organization as their employer meddling in their personal lives and family problems. According to our experiences, a very small percentage of servicemen and servicewomen are willing to talk openly about their family issues. Some of those included in research do not trust the research team regarding the anonymity of the research tools and the results, and believe everything they say will be reported to their superiors and will influence their future professional career. It is very difficult to conduct interviews amid such a mistrusting environment. For example, on several occasions servicemen were raising doubts regarding the anonymity, especially when dealing with the demographic data, stating that their superiors would be able to identify individual persons. This was the case especially among officers in peace operations, where the number of individuals of certain rank and age was, in fact, very limited. We were trying to convince them that their superiors always get only interpreted results and not rough data, so nobody would be able to identify certain individuals. However, in order to diminish the level of mistrust, we stopped asking them about their year of birth and started classifying into age groups.

Third is the question of participating in research 'by order' or voluntarily. When a research team approached servicemen and servicewomen it has always emphasized that participation in the research was voluntary, with everyone having the right to refuse participation. However, the unit chosen in the sample has usually been summoned by order of their superiors, and servicemen and servicewomen have been ordered to report to the place (barracks, classroom, etc.) where the research was taking place. Thus, despite our assertion that participation was voluntary, they had the feeling they had been ordered to participate and in some cases this may have influenced the final outcome. For example, in some cases soldiers were unwilling to participate in the research, however since they were summoned by order, they had to obey, so they expressed their unwillingness and resentments by handing back blank questionnaires.

Fourth is the question of influencing the chosen sample. In most cases, the research team did not have the final say in who would be chosen for the sample. The size of the sample and the rank of the sample (officers, non-commissioned officers [NCOs] and soldiers) were the only determinants defined by the research team. Everything else was determined inside the military organization and the research team was unable to influence who was to participate in the research in the end.

Fifth, there is the question of expressing dissatisfaction with a military job by not participating in the research. In some cases, researchers have identified that servicemen and servicewomen refused to participate in the research due to some other issues unrelated with the research. Their dissatisfaction was expressed in two ways: they refused to participate from the outset, or they participated and returned blank questionnaires or gave very simple answers. Quite often, servicemen and servicewomen were also unwilling to speak openly about the causes of their dissatisfaction which led them to refuse to participate. In one case, a research team arrived at the main base where the deployed unit had been stationed one day after a serious incident in the field had occurred. And this incident, previously unknown to the researchers, influenced the whole outcome of the fieldwork. It was only additional information and explanations by a superior commander that helped shed some light on this and assisted the researchers in interpreting the results. Which again proves the importance of triangulation of methods, since the interpretation of the results could be completely false if the additional method wasn't used.

Sixth is the question of using proper methods for appropriate ranks. Based on years of experience interviewing different ranks of servicemen and servicewomen it can be noted that not all methods are appropriate for all ranks. This is especially so with cases of using face-to-face interviews with soldiers and also some NCOs. Face-to-face interviews have proven to be a suitable method for officers, who are usually more experienced and have a higher education. This raises the question of the appropriateness of using interviews for lower ranks, whose answers are usually too simple and very similar. It is not only the use of face-to-face interviews, but also the use of group interviews that raises a dilemma. Bearing in mind the mistrusting nature of servicemen and servicewomen, it is questionable whether interviewing a group can bring added value or pose an obstacle. Some people feel very uncomfortable talking about their personal experience and problems in a group. Thus, taking this into account it can be concluded that the use of group interviews demands a thorough choice of topics to be discussed. Not all topics and issues are appropriate to be discussed in group interviews and this must be taken into account.

Seventh, there is the question of recording interviews. Experience of researching the SAF shows it is almost impossible to record conversations with members of this military organization. It is noted that recording interviews is considered undesirable since servicemen and servicewomen mostly fear that their statements might be misused later. This problem is inevitably connected with the question of trusting the researchers, fear of the research data misuse and the closed nature of military organizations. Thus, if the use of recording devices is not allowed, it is up to the researcher to write as much as one can during the interviews or try to remember as much as possible and write a report right after an interview has ended. Either way, the researcher loses one component, which is the observation of the interviewee's reactions, gestures, etc. shown on their face or in their body language. Besides that, it is also impossible to write down or remember everything the interviewee has said. Therefore, we need to accept the fact that a small amount of information will probably be lost forever. This problem might be resolved to some degree by multiplying the number of researchers conducting an individual interview. In that way, one is asking questions, observing the interviewee and their reactions, while the others are making notes. In the end, all the written records are compared and merged, yielding information of much higher quality than an interview conducted by a single researcher. However, in circumstances of financial and staff deprivation it is almost impossible to ensure a satisfactory number of researchers, let alone conduct a single interview with multiple researchers.

Eight is the question of proper timing when conducting a research in the field. According to our experiences, it can be very delicate to plan a field research considering all the different factors: number of servicemen present in the area of the mission, number of servicemen absent due to fieldwork, sick leaves or holidays. In research projects where the goal was to include all of the deployed personnel, this might present a major challenge and the deficit of an extensive part of the sample can significantly influence the final outcome of the research. For example, the most significant deficit was identified among the first (at home) and second (in the area of the mission) stage of the research among the members of the 15th Slovene contingent for KFOR. During the first stage of the research, 432 servicemen and servicewomen were included, while during the second stage 323 were included. Proper timing is also very important in regard to disturbing regular cycles of deployment. Researchers were often seen as intruders in their everyday tasks and participation in the research as 'one additional burden.' In a few cases, due to time limitations, the first stage of the research (prior to the deployment) was scheduled immediately prior to the departure, which made the participants very dissatisfied. In other cases, the third stage of the research (after the deployment) was scheduled a few weeks after the return, when servicemen and servicewomen were on leave, and were ordered to report to the barracks in order to participate in our research. The level of dissatisfaction was very high which undoubtedly manifested in the final outcome of the research.

Conclusion

Qualitative analysis is as old as the study of societies and cultures itself. It was known by different names until the second half of last century; however, it has evolved over the centuries. Qualitative analysis, as we know it today, has in a way been the same as the researches of Malinowski or Mead, yet in many ways improved. Military organization is a constituent of every society[12] and therefore is subjected to the interests of society and needs to be studied. Problems occurring within a military organization are in many aspects the same as in any other (e.g., family and friends, motivation, fears, love, trust, satisfaction with superiors, etc.). Despite that, they deserve special attention since people influenced by those problems use weapons every day they are on duty. Due to its specific features of being a closed institution with strong internal socialization, it is very hard to enter or study it. After overcoming this first barrier, the researcher needs to cope with several other obstacles and limitations narrowing the research options. It is pretty challenging to conduct research within military walls considering limitations such as the prohibition on using a tape recorder, dealing with people participating in the research because they have been ordered to do so and yet are expected not to tell too much or even reveal some inconvenient facts, feelings, ideas, observations, etc. concerning individuals. However, years of research show that certain methods are more appropriate than others. The triangulation of methods, structured and standardized open-ended interviews, combined with observation with participation and sometimes also with group interviews has proven to be the appropriate methodological approach. A structured interview has proven to be useful when asking for the opinions of the majority of servicemen and servicewomen of all ranks. For example, when trying to define what motivates SAF members in peace operations, standardized, open-ended interviews are used when conducting an in-depth analysis of certain problems, where the opinions of smaller groups are being collected – for example, opinions of only staff officers in peace operations. Group interviews have been used rarely, mostly where there was very little time and additional explanations of information gathered using another method were needed, for example, when visiting SAF members in peace operations in the field, or after the research has ended and preliminary results have been presented to the groups of servicemen and servicewomen who were the subjects of the research.

Finally, observation with or without participation has been a very useful method, especially when studying culture and other aspects of participation in peace operations. However, visits to the area of operation are, as a rule, very short (only a few days) and therefore, other methods also need to be used. Once again, the use of several different methods and a combination of different research techniques are the key to success when researching such a complex and specific institution as a military organization.

Notes

1 For example, the positivist paradigm, post-positivist paradigm, post-modern paradigm, post-structural critics of validity and reliability in qualitative analysis (Kogovšek 1998).

2 The method of observation with participation means researchers actively participate in the studied environment, and the method of observation without participation means researchers are only observing, without active participation (Bailey 1994).

3 A characteristic of the structured interview is closed questions where the interviewee must choose from a given series of answers. The researcher is not allowed to give any additional explanations during an interview nor change the questions, he/she needs to prevent a third person from disturbing the process and can never offer his/her own opinion (Kogovšek 1998: 30).

4 The researcher uses various techniques of studying data gathered with a single method.

5 One of the limitations can be gender. In certain cultures, male researchers are not allowed to talk to women and vice versa, meaning that the triangulation of researchers is needed to complete a study.

6 The person the researcher must pass to gain access to the group.

7 The concept of common people's defense and social self-protection was a concept that was valid during the period of the Socialist Federal Republic of Yugoslavia (SFRY). It was defined in 1972 by the Guidelines for the defense of the SFRY from aggression. Common people's defense was perceived as an integrative and organic function of the self-governing society for defense against aggression. This concept was an answer of the social, military and other sciences to questions concerning the defense of society (Rebolj 2008: 19).

8 The Yugoslav People's Army.

9 The complete list of research projects by the Defense Research Center of the Faculty of Social Sciences at the University of Ljubljana is much longer and is available at: www.fdv.uni-lj.si/Raziskovanje/vsak_center.asp?id=4. For the purpose of this article, only projects relevant to the debate on the appropriateness of qualitative methods for researching military organizations are mentioned.

10 The first research project started in October 2002 and lasted until August 2003. Two contingents of the SAF in the Stabilization Force (SFOR) in Bosnia and Herzegovina were included. The first project and its findings represented the basis for a further research project, which ran from October 2003 to March 2005. It included five SAF contingents, also involved in SFOR, Bosnia and Herzegovina. The third project lasted from February 2007 until August 2008 and included three Slovenian contingents in KFOR and three Slovenian contingents in the United Nations Interim Force in Lebanon (UNIFIL).

11 Based on years of researching SAF it can be, unfortunately, noted that personal ties can and do strongly influence the final outcomes of the public tenders.

12 We are ignoring those few countries that do not have military organizations.

References

Alasuutari, P. (1995) *Researching Culture: Qualitative Method and Cultural Studies*, London: Sage.

Bailey, K. (1994) *Methods of Social Research*, New York: Free Press.

Borman, K., LeCompte, M. and Preissle Goetz, J. (1986) 'Ethnographic and Qualitative Research Design and Why it Doesn't Work,' *American Behavioral Scientist*, 30(10): 42–57.

Creswell, J. (1998) *Qualitative Inquiry and Research Design*, Thousand Oaks, CA: Sage.

Dean, J., Eichhorn, R. and Dean, L. (1969) 'Limitations and Advantages of Unstructured Methods,' in G. McCall and J. Simmons (eds.) *Issues in Participant Observation*, Reading: Addison Wesley, 19–24.

Fontana, A. and Frey, J. (1994) 'Interviewing: The Art of Science,' in N. Denzin and Y. Lincoln (eds.) *Handbook of Qualitative Research*, Thousand Oaks, CA: Sage, 361–376.

Fraenkel, J. and Wallen, N. (2006) *How to Design and Evaluate Research in Education*, New York: McGraw-Hill.

Huzjan, V. (2004) 'The Qualitative Analysis of an Interview,' in J. Fikfak, F. Adam and D. Graz (eds.) *Qualitative Research*, Ljubljana: ZRC Publishing (Institute of Slovenian Ethnology at ZRC SAZU), 187–199.

Jelušič, L. and Grizold, A. (2008) *Obramboslovje: od epistemološke uresničitve do internacionalizacije (Defence Studies: from epistemological realisation to internalisation)*, Ljubljana: FDV, 189–203.

Jelušič, L. and Papler, P. (2006) 'človeški dejavnik v vojaškem sistemu' (Human factor in the military system), *Bilten Slovenske vojske*, 8(4): 7–27.

Kirk, J. and Miller, M. (1986) *Reliability and Validity in Qualitative Research*, Newbury Park, CA: Sage.

Kogovšek, T. (1998) *Kvaliteta podatkov v kvalitativnem raziskovanju (Quality of Data Gathered in Qualitative Research)*, Ljubljana: FDV (Faculty of Social Sciences).

Lincoln, Y. and Guba, E. (1985) *Naturalistic Inquiry*, Beverly Hills, CA: Sage.

Neuman, L. (1994) *Social Research Methods: Qualitative and Quantitative Approaches*, Boston: Allyn and Bacon.
Patton, M. (1990) *Qualitative Evaluation and Research Methods*, Newbury Park, CA: Sage.
Rebolj, K. (2008) *Delovanje civilne zaščite na območju občine Grosuplje med preteklostjo in prihodnjimi izzivi (Civil Protection Service in the Municipality of Grosuplje Between Past and Future Tasks)*, Ljubljana: FDV.
Sagadin, J. (1995) 'Nestandardizirani intervju,' *Sodobna pedagogika* (Journal of Contemporary Educational Studies), 46(7/8): 311–322.
Tashakkori, A. and Taddlie, C. (1998) *Mixed Methodology: Combining Qualitative and Quantitative Approaches*, Thousand Oaks; London; New Delhi: Sage.
Taylor, S. and Bogdan, R. (1984) *Introduction to Qualitative Research Methods: The Search of Meanings*, New York: John Wiley and Sons.
Vogrinc, J. (2008) *Kvalitativno raziskovanje na pedagoškem področju (Qualitative Analysis in Pedagogics)*, Ljubljana: Pedagoška Fakulteta, Univerza v Ljubljani.

9 Studying the military in a qualitative and comparative perspective: methodological challenges and issues

The example of French and German officers in European Security and Defence Policy

Delphine Deschaux-Beaume[1]

Social science, especially in France, has long ignored the analysis of the army from a sociological or political sociology perspective. There is even less methodological literature on this point, except for the significant book by Samy Cohen (Cohen 1999). This lack of analysis on the subject is significant to the extent that the military institution raises specific methodological challenges for a social researcher and requires reflexivity. The analyst actually needs to be conscious of some supposed common knowledge on the military field that must be put aside. Enquiring directly on the military ground actually shows the social scientist that the culture of secrecy in the army still exists and makes the enquiry difficult for the civilian social researcher. However, the interviewed officers and diplomats actually show a true will to communicate on their profession with the social science researchers. Consequently, the social outlines of the military field oblige us to reflect on the praxis of qualitative enquiry and more precisely on the praxis of qualitative research interview insofar as "the paradox of research interview is to have the interviewee say and show what he had until then held hidden, voluntarily or not" (Marmoz 2001: 7) by using a specific instrumentation (e.g. questionnaires, concepts) even if secrecy appears as a constitutive characteristic of the military profession and the politico-military decision-making process in France, as well as abroad.

This chapter is based on our dissertation dealing with the genesis, practices and uses of the European Security and Defence Policy (ESDP) with a focus on the comparison between France and Germany, both in the genesis and daily practices and representations of ESDP actors (military and diplomats). More precisely, we led over 130 qualitative interviews with high military officials, diplomats and political leaders in Paris, Berlin, Bonn and Brussels. Here we focus on what it means to study the military with a qualitative and comparative methodology. We will therefore raise three main issues, which are intertwined in our research. The first issue will be the qualitative perspective and the questions it raises regarding the specificity of the military mission, which is often confidential. The second issue is to concretely raise the question of the implementation of the qualitative method in the military field in a comparative perspective: how to ask questions so they make sense to the interviewees? The last issue will address the question of reflexivity, and more precisely of the position of the enquirer before the military officers. What does it mean, and how does it impact the research? We will of course rely on our case study (French and German officers in the Common Security and Defence Policy (CSDP)) to draw empirical examples, so as to illustrate the three issues raised in this chapter.

Qualitative interviewing: a methodology for a "difficult field"[2]

As for any social science method, qualitative research interviewing leads the scientist to wonder about the reasons for his methodological choices and the way the analyst constructs and collects his data. More precisely, in the case of a political sociological enquiry on the army and the politico-military milieu, qualitative interviews correspond to two main uses: getting first-hand information to the extent that most of the time the researcher does not have an extensive access to the grey literature or internal documents he would need, and having an interesting access to the military actors in a research context where the secret and very specific military language constitutes an issue for the analyst.

The interview: a socially grounded information source

In any research project on defence matters an essential methodological problem quickly emerges: the problem of access to internal documents (grey literature). In the case of European defence policy, if official European declarations are public and often available on the Internet – such as, for instance, the declarations or joint actions of the Council of the European Union, the conclusions of the EU summits, the ministerial bills of the high representatives' political discourses – the documents leading to these official papers for their part and used to prepare the official positions are actually not accessible and are protected by a strong classification system set up by the necessities of military and diplomatic confidentiality. Regarding the archives of the French Presidency of the Republic, the Ministry of Defence and the Ministry of Foreign Affairs (the Quay d'Orsay), these documents remained beyond our reach because they did not meet the 30-year notice for consultation.[3] There still exists a specific procedure to access some medium-range classified documents:[4] we therefore tried and completed this procedure at the Ministry of Defence in 2006. This authorization procedure can nonetheless turn out to be double-edged for the researcher. The higher the level of information the researcher can access, the higher the risk the research will be classified as well, which would not enable them to publish their research results in any way. And yet is it not the vocation of research in social sciences to bring to light and make understandable the research results so as to debate on them and help build a better knowledge of our contemporary societies?

We therefore opted for a simplified authorization procedure to access confidential documents. But once the authorization was obtained, it was only the beginning of a long-drawn-out process. The researcher then had to send mail to the dedicated services of the ministries of Defence and Foreign Affairs and to the Elysee Palace and the Chancellery in Paris and Berlin as well as to the archive service of the Council of the European Union where it was unfortunately given negative responses, the topic being too contemporary. This problem of accessing the documents turned out to be similar for our German field.

In addition to the contemporary characteristics of our research which tend to bar us access to most of the politico-military and diplomatic archives, the office "Studies and Documentation" of the ministerial and inter-ministerial departments of the Ministry of Defence in Paris provides a great support in understanding why our request for access to internal documents could not be met positively most of the time. The major problem concerning contemporary politico-military archives – in this case 1991 to 2007 – stems from the fact that the inventories of these documents have not yet been published and added to the archives: these internal contemporary documents are consequently dispersed among several services and departments of the institutions of defence and diplomacy, which can each refuse to communicate them.[5] The researcher found reluctance among these services' staff. To put it differently, consulting internal documents (e.g. service notices, meeting proceedings, and language elements used to elaborate national positions to be brought up on the European meeting table) depends on the goodwill of each interlocutor, who most of the time finds a justification to refuse to let the social scientist consult the documents at his disposal.[6]

Concerning the documents made accessible to a civilian researcher by the authorization procedure, they often turned out to be of little interest or unreadable owing to the use of a very opaque technical jargon.[7] Therefore, qualitative interview is the only recourse for the researcher to have access to and to understand information on how the different actors implied in the European Defence Policy decision-making process in Paris, Berlin and Brussels have come to a result or compromise. This characteristic appears to be specific to research on inaccessible social fields (Bogner and Menz 2005: 7). We consequently chose to base our research strategy and our data collection on qualitative interviews, following Howard Becker's comment that "if one wants to know society, one first has to know it first hand" (Becker 2002: 44). Samy Cohen incidentally underlines how much more fruitful than archives those interviews turn out to be in the military institution (Cohen 1999: 19): this method helps us understand and explain how the actors hold their social roles and positions and give their role its meaning (Lagroye 1997).

Direct connection with the defence field: an asset for the "profane" civilian researcher

Another, more epistemological, reason confirmed to us by our research strategy: qualitative enquiry in social sciences enables the researcher to have a direct connection to the social reality he aims at analysing (Marmoz 2001: 19). Indeed, for the civilian researcher, the defence field raises the question of the social distance with its interviewees. This social asymmetry is actually conveyed by a specific language of the defence actors' own:

> The special languages produced and reproduced by the specialist professions ... are, as every discourse, the product of a compromise between an expressive interest and a censorship constituted by the structure of the very social field within which the discourse is produced and operated.
>
> (Bourdieu 1982: 167–168)[8]

Therefore immersing oneself in the specific language and the social codes of one's interviewees turns out to be very fruitful and can only happen by repeated contact with the research field, these contacts enabling the weaving of mutual trust between the researcher and the person interviewed. Qualitative interviews constitute not only a tool for the analysis of representations (in the present case the influence of a European or pro-NATO strategy on the social representations of the French and German politico-military actors daily dealing with the construction and implementation of the Common Security and Defence Policy (CSDP) of the European Union) but this method reveals, nevertheless, the social practices of the interviewees through their discourse on their practices. The use of qualitative interviews leaves some room for the context of the discourse in the analysis. If "individuals instantly adapt their behaviour to the social scenes within which they participate in" (Beaud and Weber 2003: 334), the fact that they speak about their representations, their professional training and career and their daily practices conveys a real added-value for the researcher who is looking for the analysis of the relationship between the actor and the institution he belongs to. Jacques Lagroye for instance relies on the theory of social roles and underlines the fact that

> the relationship towards the institution is first of all the relationship between the one who holds a role in this institution.... It is first the

> apprehension of individuals living in the institution who, because they hold roles, enable us to have an idea of the institution.
>
> (Lagroye 1997: 8)

To go further on this idea, reflexivity necessitates to "set the collected discourse in the institutional context where it has been enunciated to the extent that speech can not sociologically exist independently from the institution giving it its social justification" (Bourdieu 1982: 71). This located discourse informs the researcher on the institution, its internal functioning and its lively dimension, but also outlines a bias of the methodology based on qualitative interviews: the researcher has to be careful and keep this collected discourse at a distance. The interviewees cannot be assumed to be objective as they are personally involved in the process the analyst is investigating: "often the memory of the actors is failing; they mix up dates and tend to reconstruct their role a posteriori" (Muller 2003: 94). On the side of the enquirer, the solution is to shrug off the myth of "objective truth" to replace it with "subjective and partial truths" that the researcher has to restore and confront so as to be able to make up his own supposedly disinterested point of view (Beaud and Weber 2003: 303). For us the stake was to adopt a comprehensive approach in the Weberian sense, which is inclined towards the comprehension of the internal logic of the action of our interviewees (Weber 2003). On a methodological level this scientific posture means the interviewer pushes forward the interviewee by adopting himself an "inferior" position and bearing in mind the specificity of the army based on a strong hierarchical principle (Kaufmann 2004). The researcher tries to enter the actors' world by listening to them and learning from them or even participating in their professional duty when it is materially possible, and then "getting out of the field again" when it comes to the interpretation of the collected data (Paillé and Mucchielli 2003: 229). This "going native" approach is interested in the actors' discourse and enables the analyst to access internal information if we base our research strategy on the comprehensive Weberian approach built on the following presupposition: the meaning the actor gives to his social action largely contributes to determine the formal aspects of this action. This leads us, as researchers, to pay careful attention to the meaning the military interviewees give to their action. To put it another way and freely relying on Foucault (Foucault 1969), the interview provides an access to the actors' discourse, which has to be considered as a social practice or a social event: Foucault considers discourses as a set of regulated and specific social practices among other practices.

After having clarified our research strategy which relied on qualitative interviews owing to the specificity of our research object and of the military field of enquiry, it appears important to analyse the implementation of research interviews (here semi-directed) in the defence social field. This implementation raises indeed peculiar methodological challenges particularly due to the status of military speech.

Interviewing officers in a comparative perspective: implementation and challenges

How can one carry on a qualitative research in the military social field? What are the specificities regarding the implementation of research interviews? A research approach such as ours, based on the actors, complicates the enquiry phase. In order to analyse the construction and implementation of the European defence policy, as well as the representations and practices related to this policy among the French and German actors operating it daily, the choice of the interviewees has relied on their specific social experience in this new policy. We had to reach the officers and diplomats who were working at putting on track the European defence project that is the "historical actors", whose professional rotation frequency makes them quite difficult to reach. A second type of interviewee were the officers (and also the civilians and diplomats) operating the CSDP daily, once it was launched in 1999. It then appeared essential to draw a mapping of this network of actors both in France and in Germany so as to identify the relationships of interdependence and the interactions within among their institutional positions: we had to zigzag our way up this network so as to be positioned to understand the similarities and divergences in the practices and representations within CSDP. One of the most important stakes of this method is to show the interactions between military actors located both at the national and European levels in the production of the European defence policy by relying on an in-depth study of the actors of the French–German military cooperation involved in the European defence social field. We also had to cope with distinctive military discourse collected during the interviews and limited by legal restrictions.[9]

Mapping out a social network: an insertion strategy to get round military hierarchy on the ground

The method implemented leans more generally on a threefold approach: first of all, the researcher spots the actors and their logic of action in order to discover the actors involved in the public policy sector under study (here defence). Then comes the identification of the interdependence and power flows: who is a member of the network? Who is an outsider or remains outside the network? How do the members of the network cooperate? Last but not least, the analyst examines the evolution of the institutional functioning and the impact of European integration on this process. In this matter, Andy Smith proposes to call up a range of four enquiry techniques validating this threefold approach: information collection and literature review, semi-directed interviews, participant observation (whenever possible) and budget analysis (when it is relevant) (Smith 2000: 229–252). We combined all four techniques, with a very limited use of participant observation due to the deeply rooted persistence of a culture of secrecy in the defence field. Even if the military actors often spoke more than expected a priori, participant observation is far from being welcome and facilitated.[10]

More precisely, our field enquiry relies on 135 semi-directed interviews (based on an interview grid mixing open and closed, thematic and analytical questions) led in Paris, Berlin and Brussels between 2005 and 2008 and relying on a "snowballing" technique aiming at cross-ruling the actors' networks and completing this technique by content analysis. The comparative perspective is a challenge to take into account. Thus we had to identify the institutions daily dealing with CSDP in Paris and Berlin: the difficulty was that these institutions are not completely symmetric in both states. For instance, the Elysee Palace plays a very significant role in these matters whereas the German Chancellery intervenes more selectively. Indeed in Germany the head of the armed forces is not the Chancellor but the Federal Minister of Defence, whereas in France this is the President of the Republic's role. Consequently the question for the researcher is to discover how the French and German politico-military systems work. The best way to do so appeared to be an immersion first in the Parisian defence social field, and then in the German defence social field. After an in-depth reading and content analysis of specialized literature, press and institutional websites on the subject, we therefore opted for a research stay in Berlin at the *Sozialwissenschaftliches Institut der Bundeswehr* (SOWI, or Institute of the German Army for Social Science Research). We stayed there for seven months in 2006 and five months in 2007.[11]

The interviews were led in the services listed in Table 9.1, with officers, non-commissioned officers and diplomats.

Working by interviews actually implies that if they want to collect valid data, the researcher should not be perceived as an "intruder" in the social configuration within which they enquire. This characteristic is particularly true in the military field. Second, we noticed that some contacts offer "open sesames": the practice of name-dropping was fruitful among superior officers and also at the intermediary levels.[12] For example, having met with Admiral Lanxade (a former French General chief of staff) in 2004 for a former research enabled us to meet his friend and colleague General Naumann: both held key strategic roles in building up European defence in the 1990s. The progressive integration and even the curiosity shown by some interviewees have largely been favoured by word of mouth: the politico-military system both in France and Germany, as well as in Brussels tend to function as a big family.[13] Thus one can reach the different actors, by "snowballing" from one to another. These officers dealing daily with European defence policy know one another well, they sometimes have been friends for years as some of them confessed, and have often followed the same professional training in the superior politico-military schools (Joint Forces Defence College and Institute for High National Defence

Table 9.1 Services where interviews were carried-out, 2005–2008

Paris	Ministry of Foreign Affairs • Minister's personal staff • Direction for Strategic Affairs, Security and Disarmament and notably the Under-Direction for strategic Affairs • CSFP Service	Ministry of Defence: • Minister's personal civilian and military staff • Delegation for Strategic Affairs, CSDP and NATO services • Delegation for Defence Information and Communication (DICOD) • General Military Staff: Euratlantic Division • Army Military Staff: International Relations service and Chief of Staff • Airforce Military Staff: International Relations service and Chief of Staff • Navy Military Staff: International Relations service • Interarmed Military Staff (Creil)	Elysée Palace (Presidence of the French Republic): • Personal Military Staff of the President • Diplomatic Cell (Sherpa)	Matignon and services of the Prime Minister: • General Secretariat for National Defence (SGDN) • Prime Minister's military personal staff
Berlin	Federal Ministry of Foreign Affairs: • Planunngstab (Minister's personal staff) • Politische Abteilung, Referat 202 (Direction for Political and Strategic Affairs, CSDP and CFSP Service) • Politische Abteilung, EU-KOR (Direction for Political and Strategic Affairs, EU Correspondent)	Federal Ministry of Defence: • Generalinspekteur (Chief of the General military Staff) • Presse und Informationsstab (Information Service) • FüS III-1: Bilateral Cooperation Service • FüS III-2: CSDP service • FüS III-4: International relations Service • FüS III-5: NATO service • Plannungstab (Minister's personal staff)	Chancellorship: • Abteilung 2.2.2., Gruppe 23 (Political Division) • Referat 213 (European Division)	
Brussels	• Permanent Representation of Germany to the Political and Security Comity (PSC) • French Permanent Representation to the Political and Security Committee (PSC)	• Permanent Representation of Germany to the EU Military Committee • French Permanent Representation to the EU Military Committee	General Secretariat of the EU Council: • General Direction E VIII (CSDP) • General Direction XIX (Civilian Crisis Management) • EU Military Staff	• Permanent Representation of Germany to NATO • French Permanent Representation to NATO • Supreme Headquarter Allied Powers Europe (SHAPE in Mons)

Studies in France, Federal Academy for Commandment and Federal Academy for Security Policy in Germany). Though the turnover is quite frequent,[14] a detailed study of this configuration shows that in fact the researcher often comes across the same contacts, which have evolved from one position to another in this social space.

Nonetheless qualitative interviews command that the researcher avoids the appeal of the "scoop" by relying on the technique of cross-checking which means that "a piece of information only exists if it has been given by at least two independent sources, possibly first hand" (Thoenig 1985: 40–41). As a result we opted for a multiplication of interviews at different levels of the decision-making process (from the high-ranking officers and diplomats actors down to the executing actors) so as to avoid a unilateral and official discourse and to cross the collected data and sources. This cross-checking also raises the question of the place of the speech producer in the field of this speech production (Bourdieu 1982: 170 and fol.). This place takes a particular meaning bearing in mind the hierarchy principle in defence. Often the "second knives" (Cohen 1999: 28), that is intermediary actors, turn out to be very precious interviewees: they hold a less media-related position and have few contacts with public opinion and journalists. As a matter of fact they do not internalize censorship as much as the superior officers: censorship [does not] impose its form onto [their] words" (Bourdieu 1982: 169) to the extent that they show less concern about their social image than officers of the high hierarchy who hold political positions and are most exposed to the media. Howard Becker even tends to generalize this bypassing of the hierarchy in order to study social organizations: "If we rely on the high representatives of an organization or a community to understand in detail what is going on, we systematically fail to take into account a whole range of things that this person does not hold for important" (Becker 2002: 154).

Supplementing the words of the politico-military high representatives both in Paris and Berlin as well as in Brussels by interviews with actors on the ground provides access to the actual practices and the social interactions that are developing between the lines of the European Defence Policy's orthodoxy. Moreover leading interviews with several members of the same service constitutes a way to guarantee a relatively better objectivity to the practices compared to a permanent immersion. This method implies pedagogy from the researcher, mostly towards the German interviewees: the co-signature principle (*Mitzeichnung*) in German administrative organizations supposes that every agent of a same service or division has the same information at their disposal to the extent that all the information is transmitted to everyone in the service. Therefore, multiplying the contacts with different agents in the same service was necessary to justify our request. Pedagogy also provides an efficient communication tool between the researcher and the interviewee and helps build mutual trust in the social interaction that a qualitative interview actually constitutes.

Investigating the "great mute" in comparison: a methodological challenge

Interviewing Defence personnel in France and Germany consists in carrying out an "investigation amongst a 'difficult environment', ... a suspicious environment and yet not hermetic to research" (Cohen 1999: 17). Once the contacts have been taken with our interviewees a question still remains: will the chosen interlocutors accept to talk and not just deliver a politically appropriate speech of lesser value for the researcher? Social science research in the defence social field raises an inherent dilemma for the researcher in the profound aim of his research: accessing military discourse, which is traditionally supposed to remain confidential and surrounded by secrecy, comes up against the purpose of research: disclosure and publishing of the collected data. Nonetheless, as the interviews went on, our enquiry pushed us to question a well-established prejudice: the prejudice of the army as a mute institution. In their wide majority our interviewees gave us interesting and sometimes unreleased information. Indeed as Samy Cohen observes it is of great interest for Defence personnel, officers, as well as diplomats, "not to show themselves cut from the research circles" (Cohen 1999: 17).

More fundamentally, one of the most acute problems encountered in a social science investigation on the defence environment is the status of the collected speech and its quotation. Most of the time the sources want to remain unofficial. This wish for anonymity goes hand in hand with the will to express themselves as freely as possible in front of the analyst as well as to prevent themselves from potential negative consequences of "free" expression within their office or department. Thus the officers and non-commissioned officers welcomed our questions with enthusiasm as long as we guaranteed their words would remain confidential. This actually raises the question of the "off" speech and of self-censorship: to what extent can the researcher be explicit about his sources when he investigates the military field? How can one combine research deontology and methodological rigour? This dilemma is most frequently resolved by the researchers studying the military by using the same rule as the journalists:[15] one can quote the institution and service where the interviewee works but not his name nor his function.

Moreover, in the case of a multi-country comparison as in our case (France and Germany) it is of high importance to take into account the different rules surrounding the status of military speech in each country (De Beer *et al.* 2005). The expression rules for the military agents diverge in France and Germany and strongly codify the frame of (the) authorized speech, this speech expressing a delegation of power to the member of the military institution (Bourdieu 1982: 103–119). Thus the researcher comes up against the duty of confidentiality in France. This duty has been defined by the general military status (revised in 2005) as follows: the opinion can "be expressed only outside duty and with the reserve required by the military status" (Bacchetta 2004: 76). The officers therefore have to use moderate language as long as their speech is expressed during duty: confidentiality duty does not constrain individual opinion so much as the way of expressing it. Caution is indeed much needed when the expressed opinion or the given information is to be published. In Germany however the servicemen are considered as "'citizens in uniform': officers under control and citizen[s] like the other[s]" (Pajon 2001: 245), they benefit from a rather large freedom of speech guaranteed by article 5 of the Basic Law[16] so as to protect the social connection between the army and German society. Therefore German officers sometimes publicly express their opinions against a government decision.[17] This specificity of an investigation in the defence social field also motivated the use of handheld recorders tolerated in most cases but banned on occasions.

Another challenge of comparison is to master several languages and intercultural competences. Establishing mutual trust with the interviewees is particularly important when interviewing officers and it necessitates time and a good knowledge of their mother language. As Michel Lallement and Jan Spurk observe, comparison is the "preserve of the multilingual carrier-pigeons of the [sociological] discipline" (Lallement and Spurk 2003: 71). It is also important to master the cultural codes of the countries, so as to make the interview as unremarkable as possible, even if the interview remains a specific social interaction as we analyse it below. We thus chose to lead the interviews with German officers in German rather than in English. Our research experience showed that the information given in their mother language revealed much more on their social representations and daily practices. English is actually their daily working language and leads them to use a standardized discourse. For instance, we started an interview in English with a German officer appointed in the EU Military Staff in Brussels and got no interesting information on its representations and practices. As we switched to German, the interview became much more fruitful for us and the language switch helped this officer develop more trust towards us. This example is just one out of many similar ones. Thus, without falling into a culturalist approach, we were struck by the fact that social facts – as social practices and representations of military officers are – have to be analysed by taking into account the national and organizational origins of the interviewee. Such a "sociological journey" (Gephart 2005) is a good way to develop a better comprehensive understanding of the interviewees in Paris, Berlin and Brussels.

Last but not least, as in any political science, research interviews also constitute a full social interaction between the researcher and its interviewees inasmuch as the interviewee is led to "answer for his speech" (Blanchet 1985: 113) by the way the analyst questions him. This social interaction is to be considered and analysed not only under its social dimension but also with regard to the gender of the researcher in the present case.

The researcher-agent relationship on the defence field: a gendered interaction

Qualitative interview is not only a way of getting data but also a full social interaction between the researcher and the interviewee. Analysing this social relationship actually consists in questioning the conditions of production of the "truth" expressed by the interviewed agents, military and civilian. Moreover this interaction becomes more specific when one is a young woman investigating a mostly masculine environment (Arendell 1997). Without reverting to the common cliché of traditionalism and development of stereotyped behaviours towards women, the specificity of the defence social field has an impact on the relationship between the researcher and the interviewees. Concerning our research and with regard to discussions with male colleagues also investigating this field, the gender of the analyst seems to weigh on the research interaction and indeed positively in our case.[18] Being a woman can actually be of help: it makes it possible for the researcher to ask "naïve" questions enabling her to obtain lots of information on the social practices and representations of the military and diplomatic actors. For instance the very technical aspects wrapped up in an opaque jargon were graciously explained to us unlike what happened to a male colleague asking the same kind of questions. The politico-military environment in France as well as in Germany is an exclusively masculine environment: among our interviewees fewer than 10 per cent – both officers and diplomats – were women characterized by their youth and high education level. Therefore the female researcher can sometimes feel like a curiosity and an indulgent ear for the interviewed agents of this social field. The attitude adopted towards a young woman comes close to a semi-fatherly, sometimes even a seductive range but also highlights a great concern for the image of the army within society.

Besides, any research based on qualitative interviews among the defence personnel, and more particularly among officers, has to be analysed in the light of the army–society connection both in France and Germany. The suspension of military conscription is also to be considered in that matter: if this makes the French army more reactive and able to fulfil its missions abroad in accordance with its commitments within international organizations (UE, ONU, OTAN…), this suspension also brings the military to fear the loss of the army–society connection. This connection is in France sporadically expressed by the Days of Defence Preparation and by the military parade on the French national day on 14 July. Under these conditions the researcher becomes a go-between with civil society and more precisely with the academic and scientific circles. Jean-Dominique Merchet, a French defence journalist, even evokes the need for the military to feel loved.[19] This concern for a good army–society connection also strongly exists in Germany since pacifist trends in German public opinion are still important in consequence of the former Nazi trauma. This will to communicate has to be perceived by the analyst as an incentive to cross the collected data with other sources either scientific or derived from internal documentation.

In the end investigating the defence environment and particularly interviewing military agents provides a stimulating methodological challenge which encourages the researcher to seek for inventiveness. Faced with the difficulty, and most of the time impossibility, of procuring any internal document from the military and diplomatic services, the social science researcher can only rely on the option of a quasi-immersion in the military social field. This qualitative strategy of getting the needed data, actually puts him into an investigative position endowed with some advantages: "The 'traveller' benefits from his social situation. This methodological 'outsider' is not excluded from the group but on the contrary is actually part of it" (Gephart 2005: 13). The researcher therefore has to make use of reflexivity when using the data collected in this way and to bear in mind the inherent limits to any social science research based on recurrent qualitative interviews. Conducting an interview can seem easy at first sight but still comprises limits, one of which is the possible gap between the saying and the doing of the social actors (Sala Pala and Pinson 2007). The task is to manage the risk of disconnection between these two dimensions either by direct observation when it is possible, or by multiplying interviews among the same kind of agents so as to uncover the potential discordances. Additionally, one should create a relation of mutual trust so as to reach at least partially the social practices of these actors. Indeed the researcher frequently plays the role of the scientist as outlined by Max Weber, which "obliges the individual to take into account the ultimate meaning of its own actions or at least to help him into it" (Weber 1963: 113).

In this case, if – as we have just analysed it – such a research strategy necessitates some precautions specific to the military field regarding the status of military speech and has to be based on a rigorous implementation of the research interviews, this method finds yet a good justification nevertheless through the results it enables to obtain (Lequesne 1999: 65). Keeping in mind "the island of our knowledge in the ocean of our ignorance" (Elias 1993: 124), this methodological qualitative strategy has proved fruitful to the extent that this method has enabled us to reach unreleased data on an emerging socio-political phenomenon – the Europeanization of the defence sector. That strategy has also enabled us to shed new light on the resistance of nation-states in this matter through the analysis of the practices, social representations and national policy-making processes ruling CSDP. The military institution, far from being mute, actually provides the sociologists and political scientists with a rich investigation field.

Notes

1 I would like to thank Jacques Lockwood and Scott Greer for their correcting of my chapter.
2 See Boumaza and Campana (2007).
3 In France, the bill passed on 17 July 1978 enables access to administrative documents. Nevertheless the access to the Ministry of Defence archives is restricted to people holding a security authorization procedure. The waiting time to access public archives is 30 years and even 75 years in some cases outlined by this bill of law.
4 Mostly the "restricted access" level and sometimes the "confidential" level, which are the two first scales of classification.
5 It is what happened in our case: a globally negative answer after over five months waiting. The few reachable documents belonged to an archive set taken by the high civil servant in charge after his changing of professional position.
6 This legitimate fear of the incurred risk pushed some of our interviewees to read to us some paragraphs of the documents so that we could make notes but would not show them to us.
7 Samy Cohen also states a similar assessment regarding his investigation on the Analysis and Prevision Centre of the Ministry of Foreign Affairs (Cohen 1999: 19).
8 On this point, see also Chamboredon *et al.* (1994); Cohen (1999); Laborier and Bongrand (2005: 95).
9 Here the example of the reprimand earned by the French General Vincent Desportes in July 2010 illustrates the weight of these legal restrictions on military speech. For the record General Vincent Desportes had given his personal advice on the French intervention in Afghanistan to a public media, calling this intervention an American war and criticizing the coalition strategy on the ground.
10 We have been able to carry out two direct observations: one in a parliamentary meeting at the German Bundestag on April 6th 2006; another one in a workshop organized by the German Ministry of Foreign Affairs on 18 May 2006 and bringing together German diplomats and officers around the director of the General Direction E IX of the EU Council, Claude-France Arnoult.
11 For the record, we actually really "immersed" ourselves by living two months in a garrison house of the German Army, as the SOWI is located within the former Soviet General Headquarters in Germany in Strausberg.
12 Samy Cohen states a similar assessment regarding his interviews with former French President Mitterrand (Cohen 1999: 24).
13 This affiliation feeling is increased by the professional turnover leading the individuals to maintain their institution as a professional reference point.
14 A politico-military position is actually held for two to three years.
15 It is what the defence journalists usually do. We actually asked some of them about it in January 2006: Laurent Zecchini from *Le Monde*, Arnaud De La Grange from *Le Figaro*, Jean-Dominique Merchet from *Libération* and Christian Wernicke from the *Süddeutsche Zeitung*.
16 This article states that any citizen has a right to freely express his opinion by words, writings or pictures, and the state has to watch over this freedom (Kannicht 1982).
17 It was the case for instance concerning the Bundeswehr reform by former Defence Minister Rudolf Scharping in 2000–2001, which retained conscription. See the article from Captain Jürgen Rose "Schafft endlich die Wehrpflicht ab!" (literally: "Abolish conscription!") published in *Die Welt* on 12 September 2001: military conscription does not enable Germany to fully meet its NATO and CSDP commitments regarding rapid reaction.

18 On the general question of social interaction and of the impact of gender in research interviews see Littig (2005).
19 Speech of Jean-Dominique Merchet, Monthly seminar "Young Researcher" of the former Centre for Social Science Studies on Defence now known as IRSEM, Paris, 24 January 2006.

References

Arendell, T. (1997) "Reflections on the Researcher–Researched Relationship: A Woman Interviewing Men", *Qualitative Sociology*, 2(3): 341–368.

Bacchetta, C. (2004) *Quelle liberté d'expression professionnelle pour les militaires? Enjeux et perspectives*, Institut des Hautes Etudes de Défense Nationale, Paris: Economica.

Beaud, S. and Weber, F. (2003) *Le guide de l'enquête de terrain. Produire et analyser des données ethnographiques*, Paris: La Découverte, Coll. "Repères Guides".

Becker, H. (2002) *Les ficelles du métier. Comment conduire sa recherche en sciences sociales*, Paris: La Découverte, Coll. "Repères Guides".

Blanchet, A. (1985) *L'entretien dans les sciences sociales*, Paris: Dunod.

Bogner, A. and Menz, W. (2005) "Expertenwissen und Forschungspraxis: die modernisierungstheoretische und die methodische Debatte um die Experten. Zur Einführung in ein unübersichtliches Problemfeld", in A. Bogner, B. Littig and W. Menz (eds) *Das Experteninterview, Theorie, Methode, Anwendung*, 2nd edn, Wiesbaden: VS Verlag, 7–30.

Boumaza, M. and Campana, A. (2007) "Enquêter en milieu difficile", *Revue Française de Sciences Politique*, 57(1): 5–25.

Bourdieu, P. (1982) *Ce que parler veut dire. L'économie des échanges linguistiques*, Paris: Fayard.

Chamboredon, H., Surdez, M., Pavis, F. and Willemez, L. (1994) "S'imposer aux imposants. A propos de quelques obstacles rencontrés par les sociologues débutants dans la pratique et l'usage de l'entretien", *Genèses. Sciences sociales et histoire*, 16: 114–132.

Cohen, S. (1999) "Enquêtes au sein d'un «milieu difficile»: les responsables de la politique étrangère et de defense", in S. Cohen (ed.) *L'art d'interviewer les dirigeants*, Paris: PUF, Coll. "Politique d'aujourd'hui", 17–50.

De Beer, A., Blanc, G. and Jacob, M. (2005) *L'expression professionnelle des militaires: comparaison européenne*, Paris: Centre d'Etudes en Sciences Sociales de la Défense, Coll. "Les documents du C2SD", no. 73.

Elias, N. (1993; 1st German edn: 1970) *Qu'est-ce que la sociologie?*, Paris: Pocket Agora.

Foucault, M. (1969) *L'archéologie du savoir*, Paris: Gallimard, coll. "Bibliothèque des Sciences Humaines".

Gephart, W. (2005) *Voyages sociologiques. France-Allemagne*, Paris: L'Harmattan.

Grawitz, M. and Leca, J. (1985) *Traité de science politique*, Vol. 4: *Les politiques publiques*, présenté par Jean-Claude Thoenig, Paris: PUF, p. 558.

Kannnicht, J. (1982) *Die Bundeswehr und die Medien, Material für Presse und Öffentlichkeitsarbeit in Verteidigungsfragen*, Regensburg: Walhalla U. Praetoria Verlag.

Kaufmann, J.-C. (2004) *L'entretien compréhensif*, 2nd edn, Paris: Hachette, coll. 128.

Laborier, P. and Bongrand, P. (2005) "L'entretien dans l'analyse des politiques publiques: un impensé méthodologique", *Revue Française de Science Politique*, 55(1): 73–111.

Lagroye, J. (1997) "On ne subit pas son role", *Politix*, 38: 7–17.
Lallement, M. and Spurk, J. (eds) (2003) *Stratégies de la comparaison internationale*, Paris: CNRS Editions.
Lequesne, C. (1999) "Interviewer des acteurs politico-administratifs de la construction européenne", in S. Cohen (ed.) *L'art d'interviewer les dirigeants*, Paris: PUF, Coll. "Politique d'aujourd'hui", 51–66.
Littig, B. (2005) "Interviews mit Experten und Expertinnen. Überlegung aus gechlechtertheoretischer Sicht", in A. Bogner, B. Littig and W. Menz (eds) *Das Experteninterview, Theorie, Methode, Anwendung*, 2nd edn, Wiesbaden: VS Verlag, 191–206.
Marmoz, L. (2001) "L'outil, l'objet et le sujet: les entretiens de recherche, entre le secret et la connaissance", in L. Marmoz (ed.) *L'entretien de recherche dans les sciences sociales et humaines. La place du secret*, Paris: L'Harmattan, 11–68.
Muller, P. (2003) *Les politiques publiques*, 2nd edn, Paris: PUF, Coll. "Que sais-je?".
Paillé, P. and Mucchielli, A. (2003) *L'analyse qualitative en sciences humaines et sociales*, Paris: Armand Colin, Coll. "U".
Pajon, C. (2001) *Forces armées et société dans l'Allemagne contemporaine*, Paris: L'Harmattan.
Sala Pala, V. and Pinson, G. (2007) "Peut-on vraiment se passer de l'entretien en sociologie de l'action publique?", *Revue Française de Science Politique*, 57(5): 555–598.
Smith, A. (2000) "Institutions et intégration européenne. Une méthode de recherche pour un objet problématisé", in M. Bachir, S. Duchesne *et al.*, *Les méthodes au concret, Démarches, formes de l'expérience et terrains d'investigation en science politique*, Centre Universitaire de Recherches Administratives et Politiques de Picardie, Paris: PUF, 229–252.
Weber, M. (2003; 1st German edn: 1922) *Economie et société*, Vol. 1: *Les catégories de la sociologie*, Paris: Pocket.
Weber, M. (1963; 1st French edn: 1959) *Le savant et le politique*, Paris: Editions 10/18.

10 Interviewing a group: a social dramatic art

A few remarks on dynamics and stakes of military groups

Saïd Haddad

This chapter aims to discuss the specificity of the focused or group interview technique in a military context. It deals with factors or questions which affect the research dynamics and results. Among them: the position of the researcher, including his autonomy in the selection of the interviewees and responsibility in the material organization of such interviews; the legitimacy of the groups and the data reliability; the real purpose of the interview; the accounting for non-verbal elements which shape an interview; and the political and social context where the researches take place.

This chapter is based on two surveys. The first one (referred to here as UNIFIL research)[1] was conducted in 2008–2009 and aimed to describe and to analyse the French perceptions of the United Nations Interim Forces in Lebanon (UNIFIL) as an organization and as an operation, of their counterparts (the civilian personnel of UNIFIL, the other national contingents) and the local actors such as the Lebanese population, the Lebanese Armed Forces or the Hezbollah.

This study used data from 120 questionnaires[2] collected among three units who came back from Lebanon[3] in 2007, 2008 or 2009. Nine collective interviews were also conducted in April 2008, May 2008 and June 2009. In total, there were 27 soldiers, 31 non-commissioned officers (NCOs) and 16 officers involved. In each regiment, three panels for the three ranks stood out: soldiers, NCOs and officers.[4] All the people interviewed were designated by their own regiment to represent the panels.

The second survey (Haddad *et al.* 2006) concerned the joint-services reform launched in 2004 by the French Ministry of Defense (MoD). Through the study of this reform (referred to here as joint-services research), we examined how the question of cultural diversity inside the French Armed Forces is managed. As a matter of fact, jointness brings into contact the various French services. But in spite of being legitimated by the political, strategic and operational changes, it faces some resistances due to clashes with the concerned actors' identities, practices and representations.

This research, involving semi-structured interviews, was conducted in 2004–2005 in military places, including: 26 individual interviews and one focus group. This unique focus group was composed of six Army officers, three Navy officers, three Air Force officers, one Gendarmerie officer and one from the Services. The group interview was conducted at the War College[5] where all the officers were trainees.

Both research projects underline the fact that professional and cultural diversity is experienced within the armed forces through the interservices integration or jointness process and within the multinational context during overseas operations and peacekeeping missions such as UNIFIL. Both studies also deal with the concept of culture or, to be more precise, with the cultural encountering or intercultural relations. Adopting, here, a dynamic approach, culture is 'a result of the analysis and not a given data' (Izard 2002: 191). It is all the more true that cultures exist only when confronted with one another.

Culture (and identity) 'is a matter of self-ascription and ascription by other in interaction'. So the 'critical focus of investigation from this point of view becomes the [ethnic] boundary that defines the group, not the cultural stuff that it encloses' (Barth 1998: 6 and 15).

The group interview technique

The interview technique is the keystone of both research surveys. Regardless of the type of interview (face to face or collective ones), we focused on the actors directly concerned with the topics: military from the units of the Combined Arms Battle Group deployed in South Lebanon and officers concerned by the joint-services reform. Focusing on what people told us about their experience(s) and concentrating on their perceptions (Jodelet 2003), we aimed at understanding how the personnel constructed reality. As Emile Durkheim argues, perceptions (collective or individual) are social realities (Durkheim 2007; Moscovici 2003). For instance, in South Lebanon, perceptions and representations of this UN mission varied according to the soldier's rank, the frequency of relations with others such as locals or military personnel from other armed forces, the soldier's task in his or her unit, the aim and localization of the unit, as well as the current context of the defence force. Perceptions and representations of the joint-services process and reforms varied also according to the service (the branch) the officers belong to, the importance of the services, his or her experience of multiservice operation or structure, etc. This comprehensive approach, through the comparison between the actors' self-image and their images of others', helps us to understand how French militaries make use of their cultural and material resources and how they are 'active producers of social' (Kaufman 2003: 23).

Among the differences between the two surveys was the role played by the collective interviews. If for the second research, the unique group discussion was intended to confront some hypothesis after having done other individual interviews, in the first one (the UNIFIL research), the collective interview was the condition of the survey and allowed us to go forward.

Defining a collective interview is not easy. Beyond the term, what differentiates the focus group technique from a group interview or collective one? If the aim of this chapter is not to discuss the story of focus group interviews (Duchesne and Haegel 2008: 8–34; Krueger and Casey 2009: 1–15), defining what we consider as a collective, group interview or focus group could be helpful.

A focus group can be considered as 'a special type of group in terms of purpose, size, composition and procedures' (Krueger and Casey 2009: 2). It has some distinctive characteristics, such as the number of people involved, the homogeneity of the participants, the fact that it provides qualitative data, that the discussion is focused and helps to understand the topic of interest (Krueger and Casey 2009: 6–8).

Others also define a focus group as 'a group of individuals selected and assembled by researchers to discuss and comment on, from personal experience, the topic that is the subject of the research' (Powell and Single 1996, quoted in Gibbs 1997). It means that 'group interviewing involves interviewing a number of people at the same time, the emphasis being on questions and responses between the researcher and the participants' while a focus group relies on 'interaction within the group based on topics that are supplied by the researcher' (Morgan 1997, quoted in Gibbs 1997).

L'intervention sociologique (i.e. sociological intervention) of Alain Touraine represents also another kind of group interviewing, where the feedback of all the participants (the people interviewed) is the key process of the interviews. Based on what was called 'sociology of action', the aim of this method is to act on social reality and to provide solutions.

Considering these several approaches of the focus group in the academic field as well as in the non-academic one – market or nonprofit research (Krueger and Casey 2009: 143–153) – the classical dilemma of the focus group technique concerns its purposes: is it used to obtain data produced during the interview or to observe the group and to study the interactions within it? In other words are these two aims (obtaining data and studying group interactions) not compatible with one another? In our opinion, it seems artificial to separate the collecting of data during the interview and the observation of the interactions. That's why, in order to avoid confusion, we'd rather make use of the expression 'collective interview' or 'group interviewing'.

We can't separate the interview process (discussion between the researcher and the group) from the dynamic of the group or all the interaction between all the members of the groups during the discussion. These interactions feed or supply the collected data during the interviews. The data is the result of social interactions (Duchesne and Haegel 2008: 37–40 and 42–43) or the consequences of this social dramatic art where people are obliged to commit themselves in these interactions (Goffman 1974: 102). This explains why observing how the interview is produced, how it is conducted by all the participants (the interviewers and the interviewees) and where it takes place is a part of the analysis. Interviewing and observing what is happening during the interview is a constituent of the research.

Reliability and legitimacy of the group(s)

As said above, we must notice that these collective interviews do not have the same status in the two researches. For the first one (UNIFIL), group interviewing was considered, from the start, as the first step or the exploratory stage of the study. Facing some difficulties going on the ground (in Lebanon), these interviews were used as a method in their own right and combined or completed by questionnaires and a few individual interviews. We first met soldiers, then NCOs and officers, while in the last regiment we met the officers first and the other ranks after.[6] All the interviews lasted at least two hours and were conducted inside the regiments. For the second project, the collective interview was just a complement of the individual ones, took place at the War College and occurred at the end of the data collection.

This said, what are the questions which affect the research dynamics and results?

Interviewing people designated by their own regiments to represent the panels can be considered as lacking autonomy. When we contacted the units or the War College, for the UNIFIL research or for the joint-services one, our principal demand and condition was that the interviewees be connected to the topics: having been in Lebanon under the UNIFIL II mandate or being concerned by the jointness reform. As said above, the participants were chosen by the units or by the War College.

Therefore, the lack of autonomy in the selection of the interviewees could affect the reliability of the data. Are the groups representative of the unit? Can they legitimately talk about their experiences and be the spokesmen of their unit? This lack of control in the recruitment could be prejudicial to the data quality. In fact, only people who know how to talk, who share the values of the whole regiment could be selected without guarantee that they are necessarily the most appropriate participants for the interviews. What could be the main limitation is the artificiality of the group.

But in a sense, although these interviews are organized and very formal (and not natural), the groups pre-exist. We are facing small groups (in the sociological sense) characterized by closed relationships and sharing the same specific culture (the regiment one). These groups are homogeneous, even if it was the first time – that's what we've been told – that they met together to share their experiences of UNIFIL and Lebanon. The homogeneousness of such social units is important because it means that all the participants are supposedly on equal terms. Interviews or panels per ranks (soldiers/NCOs/officers) must allow the expression of each individual view and prevent hierarchical control. Homogeneousness allows the participants to be confident during the interviewing so much so that the interview can be an unbalanced relationship (Le Breton 2008: 177) between the interviewees and the researchers or affected by a 'social dis-symmetry' (Bourdieu 1998: 1393).

However, this homogeneousness is relative: there is an explicit and implicit hierarchy or power differences inside the group. Whatever the category of rank, these differences are related to rank, of course, but also to the experience, the seniority in the army or in the unit, the individual characteristics of the participants (e.g. the good or fluent speaker). As examples, we noticed that in the NCOs groups, the sergeant majors had the monopoly of expression, while in one officers' group an older captain commissioned from the ranks was respectfully listened to by the senior officers.

Observing the interactions

The lack of control or the relatively less control over the data produced than in one-to-one interviewing can be, however, counterbalanced by the dynamics of the group discussion itself. In fact, we must notice that the multiplicity of point-of-view and even the power differences can liberate the word and can break the formal hierarchical framework which shapes the interviewing process. Through the dynamics of the group interviewing, consensus or differences of opinion can emerge and it is possible to identify individual messages during the discussion. The researchers are not facing an exclusive group, with a unique identity and a unique discourse. As Erving Goffman pointed out in *Asylums*, total identities are quite incapable of lasting even in total institutions.

As researchers, we faced here a double frame of authority: inside the group and between the group and us (see below). In other terms, the questions of artificiality and the relations of power inside the group can be combined with the questions concerning group–researcher relationships.

As mentioned before, the location of the meeting is also important and can have an influence on the interview itself. For the UNIFIL research, most of the interviewing took place in the regiments, expect for two which took place in the meeting room at the headquarters (HQ) of the unit. Despite the official character of the location (the meeting room at the HQ), the room was banal, ordinary and looked like any meeting room. We had our interviewees in front of us, on our left side and our right side.[7] There was no distance between the researcher and the participants. This seating order eases the discussion, enabling everyone to see the other express his agreement or disagreement when voicing an opinion.

This is a kind of embedding which helps to suppress the symbolic distance between the researchers and the group. To ease the interview and the voicing of personal opinions, the principle of equality between the participants has to be respected, the interview being an implicit agreement between the two sides (Hugues 1996: 16, quoted in Le Breton 2008). That's why the ordering plays an important role in suppressing the social distance between the participants and in making the interview as banal as possible. Breaking the symbolic violence or trying as much as possible to reduce it during the interview, especially with people who are considered to have a lesser social and cultural capital than the researchers (it could be the privates here) is imperative to respect this principle of equality (Bourdieu 1998).

The geography of the first interview process (in the first regiment) was a little bit different. It took place in the trophy room (or tradition room) of the regiment. This location had an impressive effect on the participants, especially the (young) soldiers and the (young) NCOs. In this case, more effort had to be made to break the social distance between the participants, especially when we were seated face to face, on both sides of the table. This order (configuration) reinforced the formal and official aspect of the interview. The exception was the officer panel, where the deputy commander took the place of the commander at the centre of the round table. So, the context of the interviewing and the status or rank of the participants play an important role during the discussion and could have an impact on the data produced.

The role of the researcher

The solemnity of the (tradition) room reinforced the feeling that the researchers were important. They were not only from the MoD and from the Military Academy of Saint Cyr but they were also allowed to use this room.

If being an insider eases the access and can facilitate the contacts with people before and during the discussion, the way we have been introduced by the authorities to the participants before our arrival could also impact the interviewing. If, for instance, we have been introduced as 'important persons' or auditors (controllers), the communication between the participants and us could be difficult (mistrust, shyness, etc.).

As social scientists working in a military academy, were we seen as (civilian) colleagues from the same institution, experts, potential spokesmen, or auditors (controllers) with a hidden agenda, as one of the interviewees suggested? Which role were our interviewees expecting from us?

Accounting for the non-verbal elements which shape an interview or the attitude during that time is not only a part of the analysis but could also help the researchers to read the roles played by the actors and the roles the participants want to assign us. As said before, the order of the meeting plays a key role in the expression of attitude, beliefs and reactions. Silence, acquiescence, pouting or shifty attitudes are part of the interview process we have to observe and to analyse. There is a permanent back and forth between the analysis of the opinions and the manner they are expressed.

That's why we must insist on the purpose of the survey: to help people feel at ease, to promote the debate and the interactions between all the participants by reducing this 'social dissymmetry' mentioned above. By moderator, we mean here a kind of mediator to gain significance from the group. But this said, the status of the interviewees has to be debated. Must we consider these militaries as informants only or have they another role? In other terms, can we consider them as full interlocutors who participate in the research?

Indeed, as the anthropologist Laplantine underlined it,

> the paradox, but also the specificity of anthropology in the field of social sciences, is not being 'the social science from the observed's point of view' (as Lévi-Strauss defines sociology), it's not either being the social science from the observer's point of view, but a practice which emerges at their borders or rather at their intersection.
>
> (Laplantine 2001: 206)

If this 'paradox' or tension is a component of the sociological or anthropological practices, we must notice that the main challenge for the researcher is to take the interviewees into account by recognizing them as full interlocutors and not as data providers only. Refusing 'the denial of the observed as interlocutors' (Chauvier 2011: 25) means, in our opinion, that the interviewees as well as the observed (who are the same persons in most of the cases here) are co-producing the research. If on the one hand, this co-production effect is the result of the mediation process which takes place inside the group, on the other hand, it's also the result of our attempt to keep the balance between our double position as insider (from the same institution and due to the adopted comprehensive approach) and as outsider (as a social scientist).

But this role of moderator also concerns all the groups of the regiments. There is a kind of dialogue between the three groups in the regiment and several perspectives appear from the groups. We also have the same process between the regiments and we can see consensus or differences of opinion emerging through the Army. In a sense, group interviewing is not independent. The people we met, discussed with and observed, form at each level what Elias called a figuration, regardless of the group dimension. This concept helps us to understand the 'interdependence chains' (Elias 1991: 159) and to have a more global view. There are explicit and implicit links or relations between the groups (inside the regiment and between the regiments): through what we have mentioned before and the questions asked by the participants interested by what their colleagues said or through the same stories or the same themes heard.

Last but not least, at an upper level – the superstructure one – the social and political context has to also be taken into account. Some fundamental elements have been shaping the social and political background of the French military organization in the last 15 years. At the top of the list is the professionalization decided by the political authorities in 1996. Since then, the military institution has experienced a lot of changes or structural reforms, among them the joint-services reorganization launched in 2004. So, if many observations and conclusions reported in our surveys are related to these structural changes, some of them are also related to the French political and strategic context. After the election of a new French president, the publication (June 2008) of a new *Livre Blanc sur la Défense et la Sécurité Nationale* (White Paper on Defence and National Security), the decision to fully integrate into NATO structures, and the change of mission for the French Forces in Afghanistan have also dramatically impacted the military establishment.

Combined with the structural reform of the Defence Forces[8] or critical budgetary tensions, these political and strategic changes emerged and shaped our interviews. For instance, the soldiers and the NCOs insisted a lot on the budget and equipment points maybe because it concerns their daily life in operation and probably also because our role and function was not very clear (researchers or officials?). Some considered us as potential spokesmen who could probably echo their opinion.

Conclusion: the interview as a field

These final remarks underline the supposed ambiguity of our position as researchers from the same institution as the interviewees. If the insiders/outsider dilemma – as written before – is finally such a classical and traditional one in our disciplines, our position as an insider from the institution has to be debated. So, the first point to discuss is the role of the researcher.

In his paper on the roles played by sociologists in the field, Raymond Gold – pursuing the works of Buford Junker – defines four possible postures for the researcher: complete participant, participant-as-observer, observer-as-participant and complete observer. The definition of an observer-as-participant may characterize our role as researchers.

In fact, observer-as-participant is defined as a role

> used in studies involving one-visit interviews. It calls for relatively more formal observation than informal observation of any kind. It also entails less risk of 'going native' than either the complete participant role or the participant role. However, because the observer-as-participant's contact with an informant is so brief, and perhaps superficial, he is more likely, than the two others, to misunderstand the informant, and to be misunderstood by him.
>
> (Gold 1958: 221)

According to this definition, the main risk is the misunderstanding between the informants and the researchers and beyond that the quality of the information secured. If being misunderstood by the informants is – sometimes – more than likely concerning our position and our function, our knowledge of the institution can counterbalance the briefness of our visit to the units. It is our key role to clarify our position and once 'a field worker has mastered his role only to the extent that he can help informants to master theirs' (Gold 1958: 222).

In addition, the second point to be debated is the interview as a field. We saw in this chapter that the interview was not only a technique but also a field we can observe. The interview is, here, the laboratory of the researcher, the place where the insider/outsider traditional dilemma emerges. The interview-as-fieldwork provides as much information as the interview-as-technique. The informants are at the same time observed and their attitude, their reaction or the way they manage the delicate relations of power inside the group, for instance, are data sources the researcher has to explore. So true is it, that 'the quality and quantity of the information secured probably depend far more upon the competence of the interviewer than upon the respondent' (Caplow, quoted in Gold 1958: 222, n7).

Notes

1 A joint research with my colleague Claude Weber. To be published. Some results are discussed in Haddad (2010). This chapter will focus on this research, except when mentioned.
2 23 officers, 65 NCOs and 32 soldiers.
3 A mechanized infantry battalion, a signal platoon and a tank company, interviews conducted in April and May 2008 and June 2009 (6th and 8th Mandates of Operation DAMAN).
4 In the first unit, we met 10 soldiers, 10 NCOs and 8 officers; in the second one: 10/10/5; in the third unit: 7/11/3).
5 At the time of the interview (April 2005), the War College (since 20 January 2011) was named Collège Interarmées de Défense.
6 Plus informal discussions during the lunch with the commanding officer, the deputy commander and the officers.
7 It was also the same ordering at the War College.
8 During our UNIFIL research, a reduction of 50,000 servicemen and women was announced.

References

Barth, F. (1998) *Ethnic Groups and Boundaries: The Social Organization of Culture Difference*, Long Grove, IL: Waveland Press.

Bourdieu, P. (1998) 'Comprendre', in P. Bourdieu (ed.) *La misère du monde*, Paris: Le Seuil.

Chauvier, E. (2011) *Anthropologie de l'ordinaire: Une conversion du regard*, Toulouse: Anacharsis.

Duchesne, S. and Haegel, F. (2008) *L'entretien collectif*, Paris: Armand Colin.
Durkheim, E. (2007) *Les règles de la méthode sociologique*, Paris: Presses Universitaires de France.
Elias, N. (1991) *Qu'est-ce que la sociologie?*, Paris: Pocket.
Gibbs, A. (1997) 'Focus Group', *Social Research Update*, 19, University of Surrey. Online. Available at: http://sru.soc.surrey.ac.uk/SRU19.html (accessed 31 August 2011).
Goffman, E. (1974) *Les rites d'interactions*, Paris: Les Editions de Minuit.
Gold, R. (1958) 'Roles in Sociological Field Observations', *Social Forces*, 36(3): 217–223.
Haddad, S. (2010) 'Teaching Diversity and Multicultural Competence to French Peacekeepers' in M. Tomforde (ed.) 'Peacekeeping and Culture', *International Peacekeeping*, 17(4): 566–577.
Haddad, S., Nogues, T. and Weber, C. (2006) *L'interarmisation: expériences vécues et représentations sociales*, Les documents du C2SD – SGA-Ministère de la Défense, No. 80, Paris: C2SD.
Hugues, E.C. (1996) *Le regard sociologique*, Paris: EHESS, p. 285 quoted in D. Le Breton (2008) *L'interactionnisme symbolique*, Paris: Presses Universitaires de France, p. 177.
Izard, M. (2002) 'Culture', in P. Bonte and M. Izard (eds) *Dictionnaire de l'ethnologie et de l'anthropologie*, Paris: Presses Universitaires de France.
Jodelet, D. (2003) 'Représentations sociales: un domaine en expansion', in D. Jodelet (ed.) *Les représentations sociales*, Paris: Presses Universitaires de France.
Kaufman, J.-P. (2003) *L'entretien compréhensif*, Paris: Nathan.
Krueger, R.A and Casey, M.A. (2009) *Focus Groups: A Practical Guide for Applied Research*, Thousand Oaks, CA: Sage.
Laplantine, F. (2001) *L'anthropologie*, Paris: Petite Bibliothèque Payot.
Le Breton, D. (2008) *L'interactionnisme symbolique*, Paris: Presses Universitaires de France.
Moscovici, P. (2003) 'Des représentations collectives aux représentations sociales: éléments pour une histoire', in D. Jodelet (ed.) *Les représentations sociales*, Paris: Presses Universitaires de France, 79–103.

11 Research on Latin America's soldiers

Generals, sergeants and guerrilla *comandantes*

Dirk Kruijt

Introduction

In this contribution, I describe the development of a research style developed while studying the Latin American military in war and in politics, and carrying out research on their adversaries: the *comandantes* of guerrilla movements; the national and local command structure of paramilitary forces; and the leadership of the legal and illegal militia members. My research is largely based on interviews with the higher echelons of the Latin American military and the Central American guerrilla leadership. I also studied the background, motives, sentiments and life histories of sergeants turned head of state or cabinet members, and I contributed twice to the oral history of the Guatemalan peace negotiations.

I always use a qualitative interview style, procuring a conversational intimacy between two persons present, at ease, and trying to hold the interest of my interlocutor by inviting him or her to merge their life and career histories with the subject at hand. I consider it the most comfortable way of following the other's way of thinking and sorting out, in a decent way, the labyrinths of his or her memory. In fact, it is interviewing on the basis of shared confidence. Maybe a better term is 'borrowed trust', confidence generated by the introduction via a dependable intermediary that guarantees that the researcher is reliable. It is my experience that, without an introduction by an intermediary, the institutional culture of silence will never be broken.

Interviewing in the sphere of confidence and trust is always a two-way communication. You have to be open about yourself, your real interests behind the interviews, rewarding private confessions with personal intimacies, letting your interlocutor meander about his or her family affairs, life experiences, specific moments of personal decision making, personal success stories, and disappointments and frustrations. The conversational tone has always to be respectful, interested, inviting, building confidence to chose between the official, ideologised or institutional history, the sometimes 'enhanced' or 'purified' version of the events, and the real and personally experienced facts.

In this contribution, I will follow my own research experiences chronologically. My research projects were, in sequential order:

- Research on the 'Revolutionary Government of the Armed Forces' of General Velasco in Peru (1968–1975), accomplished in the mid-1980s and published in the 1990s. I did my own research between 1985 and 1989, based on a meticulous revision of the existing publication and on extensive primary interviews with nearly all members of Velasco's politico-military team.[1]
- Research on the military (co-)governments in Suriname (1980–1993). My own research was carried out between 2001 and 2004 and it refers to a dictatorship of ex-sergeants turned government leaders and 'informal entrepreneurs' (coca traffickers, tropical wood exporters, gold smugglers) and a never completely documented internal war, partially along the spheres of influence of local drug and smuggling routes. The research was based on personal memoirs, private archives of former civilian politicians, ghost writers and interviews with most military leaders.[2]
- Research on the Guatemalan peace negotiations, particularly emphasising the role of two key actors, Rodrigo Asturias (the commander-in-chief of one of the three guerrilla movements) and of Julio Balconi, a general later to become the Guatemalan minister of defence. The agreements between the guerrilla and the armed forces were reached in March 1996, in Cuba, with the good offices of Fidel and Raúl Castro. This research resulted in radio and TV programmes and in two books.[3]
- Research on the military dictatorships and the guerrilla wars in Central America between the 1960s and the mid-1990s. The research was based on a thorough review of the existing literature and some 90 interviews with former guerrilla *comandantes*, former presidents and cabinet members, military commanders and peace negotiators.[4]

The Peruvian generals

The most surprising discovery I made while interviewing military men was the merging of personal life histories in the collective transformation process of a generation. In the two studies where I explicitly interviewed successive *promociones* (year groups) of military cadets (the Peruvian Velasquista officers) and the leadership generation of the three Central American guerrilla wars, I was astonished by the nearly identical career paths and empirically acquired belief systems of these two revolutionary generations. It was as if many individual life histories and career choices had been moulded by birth and social class.

During nearly two years of interviews, I was sort of 'adopted' by three of Velasco's leading team members, the generals Jorge Fernandez

Maldonado, Miguel Ángel de la Flor and Ramón Miranda. They first offered me a 'who-is-who' scratch course at the very beginning, mediated in interviews and finally brought me in contact with the secretary of the cabinet and personal lawyer of Velasco, in whose office I found, to my astonishment, a complete copy of the cabinet sessions between October 1968 and August 1975. Another fundamental source of access to confidential interviews was a good friend and colleague, Maria del Pilar Tello, the author of a collection of interviews with the Velasco generals. In most cases, I transcribed the complete text of the interviews and let the interviewees correct the first draft. In some cases, I received, in return, new manuscripts of many pages, some of them a résumé of their own diaries.

Velasco's coup in 1968 and his programme of land reform, nationalisation and mass organisations was prepared by a team of colonels who had shared many of the same youth experiences, had been children of the urban poor, had started their army career as soldiers, then corporals, then sergeants, then cadets, then officers, then colonels, then brigadiers and division generals. When I published the book with a biographical chapter about Velasco, two of his senior officers wrote a foreword, typifying the military virtues of their old commander. It could have been a military-styled CV of each of them:

> Beyond all this, we believe that the greatest value of this study is its witness to and analysis of the political personality of General Velasco: charismatic leader, of simple background, a true soldier exhibiting discipline, valor, moral rectitude, self-denial, patriotism and respect for the people.[5]

The group of officers surrounding Velasco was the military elite, trained in the spirit of military intelligence (most officers were the founders of the national intelligence system) and of the military intellectuals, the same type of nationalistic officers that went through staff school and politico-military training institutes in neighbouring Brazil. Brazilian and Peruvian military intellectuals, of different ideological stances, kept key functions in the general staff, at the higher military academies, and within the (military or national) intelligence services. Planning, administrative experience and governmental responsibilities loom large in the curricula for colonels and brigadiers. From here to the formulation of regional and national security theses and to national development plans is not a tremendous step. Another step is that from formulation to implementation.

The entire Velasquista generation was educated with the concepts of the Peruvian socialist-indigenista authors like Mariateguí in the 1930s, with romantic references to the protective, Spartan, communalist Inca Empire of the past, destructed by Spanish colonialism and imperialism. Underdevelopment was explained in terms of the dependency theory, elaborated at the UN offices in Santiago de Chile in the late 1960s. Some officers were influenced as well by the liberation theology that sought to bridge the gap between the Bible books and the philosophy of the young Marx. At the Centro de Altos Estudios Militares (CAEM), founded in the early 1960s, many of the young majors and junior colonels voiced their ideas about the need for national development through 'progressive forces'. As one member of this generation expressed it:

> We have to be an army the objectives of which can also be measured by how many miles of roads have been built, by how many thousands of acres are added to the arable, the number of individuals who learned to read or write, the stretch of irrigation canals, the number of rural health clinics, how many zones were brought under national control. That is to say, [we have to be] an army that is a *symbol* for all those countries which, like our own, are in a phase of underdevelopment, hampered by the lack of capital, trained and specialised labour and faced with an *enormous amount of work*, frustrated by an egoistic and socially unfeeling leading class and a people bereft of faith, motivation and hope, collapsed and almost terminally ill with the effects of deception and exploitation. ... [This country needs] officers who, like the crusaders of old, are sparked by the fires of *credo* and *mystique* to serve not just the army, but to contribute to the development of the nation as well.[6]

The 11 members of the team of Velasco were colonels on the day of the coup in October 1968. All belonged to the military intelligentsia. In 1967, Velasco became the army commander and started to form his personal staff around him. Four of these colonels were to write the future government Plan INCA. Two others drafted the military operations for the coup. All 11 were appointed, either on the day of the coup or very soon thereafter, to the newly created Presidential Advisory Committee (COAP), the presidential think tank. It shared and sometimes moulded his ideas, jointly acted as his political memory, advised him, criticised him (not many around Velasco dared to do so) and were involved in all the important reforms and matters of state.

Many of them first became common soldiers, like Velasco himself, to bring home some earnings. Most were born in small villages in the interior. Theirs was the long road to officer training, via promotions to corporal and sergeant second and first class. All of them knew what poverty was and how it was to feel hungry. In the years between the mid-1930s and the mid-1940s the armed forces could recruit their cadre and officers from the flower of the nation. On account of the political situation, most of the public universities, gateways to the academic world and filter for social mobility from the lower middle classes, were closed. The private universities charged sizable tuition fees and the children of the rich could always get an education abroad. The number of candidates for officers' training rose from 30 to 300 per year. The very best of an entire national generation enrolled in military schools.

They became friends, dated the same kind of girls, married in the same years and sometimes became brother-in-laws. Friendships, family bonds and symbolic kinship (*compadrazgo* relations) were formed for a lifetime. In many cases, they were each other's substitutes or deputies or successors. They came to share a number of core views and values. They read the same books, were influenced by the same authors, by the same religious leaders with affinity to the liberation theology (although nearly all were agnostic). Their political ideas were vaguely socialist. Their ethno-historical views were formed by the 'socialist empire of the Incas'. Their instructors at the military schools had been first-class scholarly progressive and nationalistic authors. Socialist authors like José Maria Argüedas and Carlos Mariateguí, founder of the Socialist Party in the 1930s, were their favourites. They had graduated at the top of their class or with honours. They had entered staff schools, as students first, as instructors next. Nearly all were involved in the reorganisation of the armed forces in the 1950s. They circulated among staff functions, teaching positions and the intelligence services. In the mid-1960s, they started internal study groups. They met in Lima on fixed dates, to have a drink together, but also to discuss politics, the national situation, the problem of integrating Peru, and so on.

Throughout the 1960s, their nationalistic and progressive orientation continued, even when most of them were working as intelligence officers or battle group commanders in counterinsurgency operations against the guerrillas. In 1965, three guerrilla movements took form in the north, in the centre and in the south of the Andes region. Most of the members of the Velasco team had been commissioned to the guerrilla regions in the indigenous highlands and were shocked by the inequality, the illiteracy and the injustice, the background of their indigenous recruits. At headquarters, they received the junior officers who came to report. They heard about the working conditions, the pittance the peasants were paid, the fact that the peasants were invariably allotted the poorer land which skirted the hills. They were given some seed and had to turn in part of the harvest. Their land was full of rocks. The landlords went to extremes, prohibiting schools, for instance. These things were discussed by the incoming officers who were indignant. They were nauseated when they drafted their reports. They complained against their superiors:

> Yes, we are putting down the guerrillas but we are forgetting something. We are forgetting what causes guerrilla in the first place. We put an end to the effect, leaving the cause untouched. And it is the cause that is to be removed; otherwise it will begin over and over again!

And so it was that they began to think that to remove the causes you need change, structural change, by structural reforms. And that, when no political party will plan these reforms, the armed forces have to do it. It took only a couple of months to convince themselves and to make it clear to their commanding officer that Peru needed a coup and a revolutionary government to overcome the structure of injustice and to restructure a society that needed growth and development for all.

The Surinamese sergeants

Suriname, a former colony of the Netherlands, has been governed as an 'autonomous country' within the Kingdom since 1954 and became an independent state in 1975. Tax levies on bauxite exports, development aid and euro remittances are the country's principal (legal) income.[7] During the negotiations about independence, the Dutch government agreed on an aid package of 1.6 billion. Between 2000 and 2002, I was one of two research leaders in charge of a large-scale evaluation of the post-colonial relations and the effects of this assistance package. The report, a disaster study about neo-colonial relations and the waste of public money, was only published two years thereafter, under explicit pressure of the (Dutch) parliament.[8] It resulted in good personal relations with a segment of the Surinamese political elite.

The (then civilian) Surinamese government had also demanded a national army in 1975; the Dutch sent back a few colonial officers and many more corporals and sergeants of Surinamese descent. In February 1980, less than five years after independence, a group of sergeants waged a coup. Sergeant-major Desi Bouterse, quickly promoted to lieutenant colonel, was to play a vital role on the political stage for the next 30 years. Between 1980 and 1993, he was commander of the forces with dictatorial power in politics. In 1993, he had to resign from his military career, but as 'serial' president during the 1980s and backstage leader from the end of the 1980s to the mid-1990s, as 'national advisor' in the second half of the 1990s and as elected politician and party leader afterwards, he has been an important player in the political arena. In 2010, he was elected president of Suriname.

When my colleague Wim Hoogbergen and I decided to write the history of the military governments and the never-documented history of an unnamed civil war in Eastern Suriname, we could fall back on the civilian politicians who had been ousted by Bouterse. They provided us with their diaries, with documentation of the military operations and they granted us long interviews, even making arrangements with other politicians. Interviewing the former sergeants was not very difficult. Some of them were socialists and were afterwards jailed by Bouterse. Some joined Bouterse in profitable business affairs. Others were killed in action or committed assisted 'suicide'; they could conveniently be blamed for crimes later discovered. Those who survived were eager to recount their version of the 'revolution', sometimes in collective interviews in the presence of 'Gestapo', Bouterse's handy man for wet jobs, in later years a graduate of the Brazilian Escola Superior de Guerra and Suriname's roving ambassador in Argentina, Uruguay and Brazil. The only person we could not convince to be interviewed was Bouterse himself. His lawyers had advised him not to discuss with 'colonial researchers'.[9] We then fell back to another 'colonial' strategy: we used, by intermediaries, the same sources that, during the military governments, had provided the Dutch embassy with precise and reliable information from within the military advisory group, not knowing the identity of the source but appreciating the detailed inside knowledge.

Between 1980 and 1987, seven successive civil–military governments governed Suriname. In the same period the economy slowly collapsed due to a severe world recession in the bauxite industry, the cessation of the Dutch development aid in 1982 and the policy of monetary financing by the military cabinets. Along with its economic downfall, Suriname was plagued by civil war (1986–1992). As a result of this war, thousands of Maroons migrated to French Guyana.[10] In Suriname, during the decade of military rule in the 1980s, a parallel economy based on drug trafficking, gold exploitation and timber export amounted to 40–60 per cent of the official national economy. This economy financed the civil war between 1986 and 1992, when the regular army, insurgent Maroon groups and several paramilitary forces were engaged in guerrilla and counter-insurgency campaigns. At the same time, the military leaders on both sides engaged in cocaine partnerships during the truces and peace negotiations.

In December 1982, probably on the direct order of army leader Bouterse, fifteen opponents of the administration were executed without any form of trial. After these 'December murders', military rule became more and more dictatorial. Bouterse himself took over power from a civil council with military advisors. Meanwhile a large number of Maroons had joined the National Army and among the recruits was an intelligent soldier by the name of Ronnie Brunswijk. He quickly carved out a career in the army. Bouterse sent him to a special commando training unit on Cuba and then installed him as a member of his private security service.[11] But Brunswijk was fired in 1984. From August 1985, Brunswijk led a group of other young Maroons attacking banks and military vehicles, gaining himself the reputation of a modern-day Robin Hood. The Surinamese leaders of anti-Bouterse groups in the Netherlands wanted to meet with him. Brunswijk travelled to Europe and on his return to Suriname found himself appointed as leader of the 'Surinamese National Liberation Army' or 'Jungle Commando', and was given all kinds of promises regarding financial support, most of which came to nothing.

The acts of resistance escalated quickly. In an attempt to isolate the Jungle Commando from its sympathisers, the Surinamese army resorted to intimidating the entire Maroon population in the Moengo and Maroni River area. Soon the Maroons were the victims of counter-insurgency operations. Villages and settlements were plundered, burned down and flattened with bulldozers. Sometimes the military shot down everyone in their sight, killing scores of people, including pregnant women and small children. The death toll, according to OAS observers, came to 300 people. Almost 10,000 Surinamese refugees (approximately 8,500 Maroons and about 1,500 others, mainly indigenous people) fled to French Guyana.

In an 'emptied' East Suriname, the Jungle Commando tried to wreak revenge for what had been done to the Maroons. This was no longer limited to attacks on military targets, but specifically on economic targets. The result of the various attacks was that the Surinamese military made contact with the politicians who had been ousted in 1980. The elections in 1987 returned the old civilian parties to power, but army leader Bouterse had managed to force so many concessions from the 'old' politicians that he was able to put his stamp on the developments in Suriname for many more years to come.

In 1988, the intensity of the fighting tapered off. In July 1989, an accord was reached in Kourou (the European rocket base in French Guyana). The military, however, sabotaged the peace plans. The civilian government was strongly intimidated by Bouterse; most cabinet members were never informed about the whereabouts of the internal war. Bouterse was the commander-in-chief and nobody wanted to provoke the commander's irritation. The National Army began arming some indigenous groups and gained political influence in the interior, while at the same time remaining untouchable. After the failure of the Kourou accord and the re-emergence of the civil war, long and complicated negotiations lingered between the government and two, then three, then four, and then five armed actors: the armed forces, the Jungle Commando and several smaller paramilitary units supported by the military. In 1992, a more definite peace accord (the Accord on National Reconciliation and Development) was signed under the supervision of the OAS.

In 1990, Bouterse staged a second coup, by telephone. His second-in-command 'advised' the civilian government to leave the presidential palace; otherwise, they would be shot out. Under international protest, a 'civilian' interim government was formed to organise elections. The old party alliance won the elections again. Bouterse preserved his position as army chief. In fact he had succeeded in a partition of the Surinamese territory between his military friends and his former adversary Brunswijk. Bouterse and Brunswijk remained in control of particular corridors and transition routes for drugs, gold, tropical hardwood and luxury articles.

Six years of war left the interior of Suriname in a state of collapse. Peace was restored, but reconstruction has never really come about. As a result of the war East Suriname became almost totally isolated from the rest of the country. The trade routes from the interior to the coast moved to French Guyana and food and other provisions had to be paid for in French francs, then in euros. The Jungle Commando stimulated gold mining and smuggling as a source of income for warfare. Drug trafficking became another source of income. Gold replaced hard currency in East Suriname. Since that time, a decigram of gold has represented a particular value and everyone knows the equivalent euro or dollar rate.

In 1993, Bouterse's economic and military activities produced another conflict with the civilian government. The army commander offered his resignation, threatening to stage 'another coup by the boys who felt insulted'. This time, the junior officers, disgusted by the war years and the privileges and extra-legal business careers of their superiors, formed a protective cadre around the government buildings and asked Bouterse publicly to step down. Their leader, mayor Mercuur, was promoted to colonel and appointed as the new army commander. Bouterse, always streetwise and this time well-advised, entered the political arena, already a wealthy businessman, as a civilian politician and a new-born Christian. He participated in the national elections of 1995, 2000 and 2005, continuously acquiring more votes and more prestige. In 2010, he formed a political alliance with his old adversaries in which politician and congress member Brunswijk participated as well. Bouterse was elected president and Brunswijk is in charge of national intelligence.

The Guatemalan peace negotiations

In 1998 and 1999, I was a member of a radio and TV documentary team of the Dutch public channel VPRO. We made two large documentaries, focusing on two key persons, who during the official peace negotiations had initiated private conversations and slowly developed a kind of personal trust, even friendship: guerrilla leader Rodrigo Asturias and army leader general Julio Balconi. We interviewed both protagonists extensively during three weeks. We used only an hour of the 30-hour-plus interviews but afterwards we organised the transcriptions of the remaining tapes into chapters of a book that sold out in a month. Julio Balconi was not completely happy with the way he was portrayed. He felt that we had failed to research the exact circumstances of the army–guerrilla negotiations, and after a while, we decided to start a new book project, based on Balconi's (very good) memory, his diaries and the recently published documentation about the peace process.

The army–guerrilla peace talks had always been surrounded by a culture of silence. The first informal and off-the-record contact took place in hotels in Costa Rica, with the then civilian president-elect Vinicio Cerezo. There were later informal tête-à-têtes at El Escorial in Spain, where the Spanish crown hosted talks between guerrilla representatives and civilian and military delegates. The head of the delegation and a close friend of Cerezo, was assassinated upon his return to Guatemala 'under mysterious circumstances' – the standard expression in those days for politically motivated killings. Cerezo created a National Reconciliation Commission (CNR) headed by the archbishop (and, later, cardinal) of Guatemala; other members were politicians of the various parties, with a retired military politician serving as liaison between the guerrilla and the generals.

The old guard of hardliners in the Guatemalan military, who were still in power, was extremely reluctant to engage in peace negotiations. But a new president (Serrano Elías), elected in 1991, decided to take on the military establishment and, as supreme commander, explicitly instructed the recalcitrant generals to form a permanent military mission that would serve as an integral part of the government delegation. As luck would have it, the senior colonels and junior generals appointed to this mission were soon thereafter promoted to high-level positions in the military hierarchy. This new generation of officers was finally convinced of the necessity of a negotiated end of the war. In military terms, the prevalence of the army over the guerrilla forces was evident. The Army Command wanted to continue fighting but gradually became convinced of the necessity of a negotiated settlement, provided that the important gains that they had achieved would be confirmed as part of any final accord. The Guatemalan guerrilla leadership was prepared to continue fighting for another decade or more in the event that peace talks did not lead to a satisfactory outcome. It was under the burden of these historically laden expectations that negotiations proceeded fitfully between the years of 1991 and 1996.

President Serrano opted, in May 1993, to suspend the constitution and to carry out a self-coup with the assistance of the army. In Guatemala, however, this coup attempt was immediately greeted by widespread popular protest. The army leadership, which in turn began to waver in its resolve, consulted the constitutional court, which declared the attempted overthrow unconstitutional. After the failed self-coup, the Guatemalan parliament appointed De León Carpio, at that time the nation's Human Rights ombudsman, as president. Shortly after taking office, De León Carpio purged the military hardliners from his cabinet and appointed instead more progressive ministers who were much more acceptable negotiating partners in the eyes of the guerrillas. A newly formed Asamblea de la Sociedad Civil (ASC), headed by the archbishop and consisting of labour union officials, journalists, Maya representatives and leaders of other popular movements, functioned as a kind of extra-parliamentary support group. In 1994, as a consequence of the first partial accord – concerning human rights – a UN verification mission (MINUGUA) was established in Guatemala: its mission, following the final peace agreement in 1996, was to monitor compliance with the terms of the settlement.

In April 1991, the first in a long series of peace talks took place in Mexico. The government negotiating team comprised civilian government appointees, along with four military delegates. The guerrillas were represented by the senior *comandantes*, accompanied by their advisers. The formal negotiations carried on over the course of six years, during which time there were several interruptions when the guerrilla leadership thought it better to retire from the negotiating table. In December 1996, the final peace agreement was signed, making effective all previous partial accords. It is fair to conclude that the success of the security-related agreements was the result of a gradual rapprochement between the army and the guerrillas. The personalities of the two main protagonists, Asturias and Balconi, were decisive. During lunch or dinner breaks in the course of each two-week negotiating session, they developed a strong rapport and mutual trust. At one point, each man agreed to keep the other informed about sensitivities and susceptibilities within his own camp in order to avoid unnecessary friction during the public sessions. In private conversations they exchanged impressions about how proposals on demobilisation and disarmament, the reduction of the Army, the abolishing of the various police organisations, the new security doctrines, and other matters, would be received by the other side. In early 1993, they worked together to establish a direct dialogue between the Army Command and the guerrilla leadership. After initial sessions between Balconi and the four guerrilla *comandantes*, the army command convinced the president of the convenience of high-level discussions between an army delegation and the guerrilla leadership. The first session was held with little fanfare in Cancun, Mexico.

When in 1996 newly elected President Arzú appointed Balconi as minister of defence, negotiations between army and guerrilla leadership were intensified – with the full consent of the president. They asked each other, in confidence, 'What will happen to us after the peace?' At another session, the guerrillas suggested, half-jokingly, that the next session should be held in Cuba. Balconi accepted the challenge and organised – through intermediaries – a three-day session in Havana, under the auspices of Fidel and Raúl Castro. The Havana session marked the decisive reconciliation between the army and the guerrillas. Immediately afterwards, the *comandantes* announced a unilateral cease-fire and Balconi ordered the disbanding of the hated paramilitary patrols. The army staff and the second-in-commands of the guerrillas worked out the timetable of disarmament of the guerrilla forces and the peace was signed in December 1996.

We produced the book based on interviews during three weeks, three times per day, seven days per week: between 10.00 and 12.00 hours, between 14.30 and 17.00 hours and between 19.00 and 21.00 hours. Military men are Spartans by heart and both Balconi and I wished to make an uninterrupted series of working sessions. Previously, we had made an outline of a possible chapter series, itemising a short history of the Guatemalan armed forces, its internal structure and the intelligence (at that time) about their adversaries. We also decided to insert a personal history, a career history and a general history about the war decades. At the end of each interview day, we decided upon the points of interest for the next two days. Probably General Balconi was more disciplined than I: he always had worked out his notes based on his diaries the very morning after our last evening session. His daughter, a political scientist, transcribed the interviews and sometimes corrected factual data. Afterwards, I grouped the data of the 60 interviews in draft chapters. Balconi then rewrote part of the text (I would have preferred the original taped version with its vernacular and side comments, but it was and is the General's book). In Cuba, we reviewed the final text a year after the interviews, when General Balconi was convalescing from a nearly mortal accident (or assault).

The Central American guerrilla *comandantes*

Interviewing the guerrilla leadership of El Salvador, Guatemala and Nicaragua meant rediscovering the strange relationship between personal life histories and the evolution of a revolutionary generation. Between 1988 and 1992, I worked as a development diplomat in Central America, spending a great deal of time in El Salvador, Guatemala and Nicaragua. During these years, I became acquainted with the personalities and institutions directly involved in the peace negotiations and the post-war reintegration process. Three parallel wars were fought by three guerrilla organisations, each of which was led by a generation of young urban intellectuals. They aspired to overthrow military dictatorships in their countries and to establish socialist societies that would root out the corruption and inequality of the previous dictatorships and oligarchies that had long held political and economic power.

Primary data comprised more than 90 interviews with political and military leaders on both sides of each conflict, as well as with social scientists and intellectuals. Most of these interviews were conducted between 2004 and 2007. In Guatemala, rounds of interviews were also carried out in 1994 and in 1999.[12] Nearly all interviews were conducted as open-ended conversations that lasted between one and three hours, with a number of individuals being interviewed more than once. In the case of Rodrigo Asturias and Julio Balconi (see the above paragraph) these interviews took place over a period of several years. They were of great help when I started the research interviews in Guatemala. Afterwards, I spent a couple of months in El Salvador, where other colleagues were very generous with the introductions. Ruben Zamora, after the Salvadoran peace agreements the first presidential candidate on behalf of the Frente, provided me free access to his private archive, his own interviews with the guerrilla generation and contacted me with other prominent guerrilla members. In

Nicaragua, I interviewed in three rounds in 2006, five months in total. Some of the much respected *comandantes* made introductions to their colleagues. Margarita Vannini, director general of the Instituto de Historia de Nicaragua and Central America at the Universidad Centroamericana of Managua (the former Instituto de Estudios Sandinistas), one of the cultural meeting places in Nicaragua, convinced other key persons to let me interview them. It was only through her personal intervention that I was able to conduct a number of sensitive interviews. That proves the fact that one can only start working through the generous assistance of respected intermediaries and colleagues.

Sometimes, the respondents also provided me their written reflections in the form of diaries, personal documents or formal memoirs. Products of the literary imagination in each of these three countries were also consulted; Central America is the region of soldier-novelists and hero-poets.[13] Several peace negotiators and cabinet members provided me with access to their private archives. I also benefited from listening to the taped interviews and reading the written sources that previous researchers had used in their publications and which are housed in a special archive at the library of the Universidad Centroamericana of San Salvador. I made extensive use of primary sources, including lengthy and in-depth interviews, unpublished private archives, scholarly studies and government reports, sympathetic and adversarial position papers and white books, published and unpublished memoirs, mimeographs and booklets distributed by the guerrilla movements. In addition, numerous secondary sources were consulted, including interviews conducted by other researchers and journalists.[14]

As in the case of the Velasco generals in Peru, a politically restless generation emerged in El Salvador, Guatemala and Nicaragua, strongly influenced by feelings of outrage about dictatorship and social injustice. This time, the members of the revolutionary generations were mostly children of the urban middle-class families that studied at the national and the Jesuit universities. The late 1960s and early 1970s were the heydays of the dependency theory and the liberation theology. Many of the young people of this generation found an outlet for their discontent with the established order in student movements at secondary educational institutes and universities, and many of these students were recruited by the guerrilla movements.

Another revolutionary cohort emerged from the religiously inspired Christian Base Communities, which supplied much of the leadership of the Central American popular organisations in the 1960s and 1970s. These groups drew ideological sustenance from the liberation theology, which also gained significant influence in student movements. Probably half of the Nicaraguan and Salvadoran *comandantes* had been associated with both the Catholic study groups. A smaller number of leaders were drawn from other sources: trade unions, grassroots organisations in the urban barrios, and peasant associations. The cadres of the left-wing parties, mainly the Communist Party, also provided prominent revolutionary leaders in Guatemala and El Salvador. Several small politico-military organisations, the constituent parts of the later national guerrilla organisations, were founded by ex-members of the radicalised Communist youth.

Finally, young military officers sometimes changed sides and joined the underground with veteran officers serving as military instructors of young guerrilla recruits. Before the Salvadoran and the Guatemalan armies were transformed into counter-insurgency machines, a certain proclivity to rebellion had existed among the young cadets and junior officers from the mid-1940s to the late 1950s. The Nicaraguan and Salvadoran *comandantes* were very young. The Sandinista guerrilla leaders were around age 25 when they defeated the Somoza regime in 1979; the only exception was middle-aged Tomás Borge. Most of the Salvadoran leadership was even younger.

Santiago Santa Cruz, first a doctor and then a *comandante* of a guerrilla battle group, describes the situation at the Guatemala universities in the late 1970s:

> You had to do your residency in public hospitals, where extreme poverty and gross inequality were always evident. In addition, many of the professors were active in the ranks of ORPA [the guerrilla organisation of Rodrigo Asturias]. This was true in the case of the [coordinator of the] program of the supervised professional internship, our rural residency program. There were other doctors who were also active. Thus, the entire infrastructure of the Medical School of the San Carlos University was put at the disposal of ORPA in order to supply its battle groups – the cars, the land, the school were on. Many doctors who were sympathetic to the cause and who came to supervise the students in the different towns out there carried arms and ammunition, and transported other material needed by the guerrillas.[15]

Of the 60 interviewed guerrilla *comandantes*, many explicitly mentioned their university years with nostalgia. At least 20 of them returned afterwards to academia or to research NGOs. Some of them were even invited to US or European universities to lecture about post-war politics. Even during the war years, several of the Salvadoran key *comandantes* found the time to publish in academic journals or to contribute chapters in academic publications.

The influence of religion was also enormous. Contrary to what one might think from hearing the incessant railing of the Reagan administration against godless and wicked communism in Central America, the reality was that Catholic priests, as well as ministers and lay-persons from various Protestant denominations, entered the ranks of the different guerrilla movements in considerable numbers. One of them is Salvadoran *comandante* Dagoberto Gutiérrez:

> I received a Christian education. But I am much more of a communist than I was before, and I'm more than 60 years old now, you know. Because at that time I didn't know anything about politics, except what I had learned from Jesus Christ. And I knew nothing about communism and capitalism. I learned about that from Marx. Afterward, I discovered an affinity between Jesus and Marx.... If you look at those of us who led the war, you see university graduates, professionals, priests. This marriage between the resistance and the churches was inevitable. It was a wedding that had been foretold. Why was this the case? Our movement was never either anticlerical or atheistic. Never. Even the Communist Party had Bible-toting priests in its ranks. We communists had our own pastors to explain the revolution. There was no anti-clericalism, no atheism.... And I'm talking about all of the churches. That relationship is very important – it was ecumenical. Both the Catholic and Protestant churches were committed. There were never any wars here between the two parties. The clueless gringos never knew who they were fighting.[16]

And then there was the influence of Marxism, a special form of Marxism, that of 'communism with a Cuban face'. The influence of the Cuban revolution and the Cuban example were without a doubt irresistible. Cuba and two of its protagonists, Fidel Castro and Che Guevara, towered above all other personalities and movements in terms of importance. Nearly all Central American *comandantes* had been in Cuba more than once for one reason or another: military training, consultation with the Cuban leaders, medical treatment, study, rest and relaxation. With a few exceptions, the only victorious post-revolutionary society they knew from direct experience was Cuban socialism. The guerrilla campaign with Fidel as strategist and Che Guevara as campaign hero, along with Raúl Castro's rapid forging of the Fuerzas Armadas Revolucionarias into a formidable military force, generated a profound respect for Cuba among successive generations of leftists-nationalists throughout Latin America. Castro's routing of US trained forces at the Bay of Pigs confirmed his iconic status. Guevara was a prolific writer who published extensively on guerrilla theory and who became the venerated guru of insurrectionist strategy. The high regard of Latin America's revolutionaries for the Cuban leadership meant that the Cuban model served for two decades as the living image of an ideal society for every leftist-nationalist insurgent. Being a revolutionary and identifying yourself as Marxist-Leninist thus came to be seen as one and the same thing.

Cuba's support of the Central American guerrilleros was, though steadfast and continuous, basically restricted to providing military training and political and military strategic advice. Given the age and the nearly legendary status of Fidel Castro, he assumed something of a fatherly role vis-à-vis the Central American *comandantes.* Che Guevara's writings on guerrilla strategy, his personal record in guerrilla combat and his heroic death after being captured in battle, along with his uncompromising dedication to the revolutionary cause, made him something of a civil saint among Latin American revolutionaries. Nearly all the *comandantes* that I interviewed acknowledged the importance of Fidel and Che as role models. In the 1970s, young Sandinista recruits in Nicaragua solemnly declared their loyalty 'before Fatherland, history, and Che Guevara'. Fatherly Fidel would remain an icon throughout the decades of the Central American wars.

After the Sandinista revolution in 1979, Managua and La Habana became the rest and relaxation places of the Central American revolutionaries. Between the Nicaraguan and Salvadoran guerrilla leaders of the same age, sharing the same values and fighting experiences, long-lasting friendships developed until the present. After the early 1980s, the Guatemalan guerrilla generation, at the defensive and mostly operating in remote rural areas (whereas the leading *comandantes* stayed in Mexico City) lost contact with their Central American and Cuban counterparts. When they returned to Cuba during the army–guerrilla negotiations in early 1996, it was – significantly – the Guatemalan army that had made the contacts through intermediaries with the Cuban leadership.

However, fifteen years after the peace agreements in Guatemala, twenty years after the peace in El Salvador and more than 30 years after the overthrow of the dictatorship of Somoza in Nicaragua, the guerrilla generations still exist as 'generation'. With a few exceptions, they use their guerrilla ranks in private conversations. In El Salvador and Guatemala they reconciled with their former military adversaries. Between them there is a kind of institutional respect; they always use the correct rank and formalities. It is the soldierly profession that bridges the ideological distance. In Nicaragua, the former guerrilla officers still form the backbone of the two strongest and most respected national institutions: the armed forces and the national police.

Methodological considerations

This style of interviewing is the topical instrument for the oral history of implicit and explicit strategies and tactics during civil wars and revolutions, including military revolutions and military governments; about political and military decisions within the inner circle; about orthodoxy, dominant ideas and heterodox discussions within the leadership; about covert actions and hidden negotiations. Here it refers, in fact, to all ingredients needed to accomplish the critically needed information and assessment of vital non-official contemporary history: counter-insurgency operations, peace negotiations, agreements between the military and 'civilian' politicians, pre-coup bargaining between economic and political elites and posterior military governments.

Performing research on Latin American soldiers, be it military personnel, guerrilla members or affiliates of paramilitary organisations, requires sensitivity about the military ethos and culture, *esprit de corps* and a culture of silence vis-à-vis 'civilians'. I use the term 'civilians' referring to two distinctive segments: the general public as explicitly distinct from military and security affairs, and the sort of 'extended family civilians' of military and security significance: diplomats, historians, social and political scientists. Another variety are the 'professionals' and 'technicians' like doctors and engineers, generally incorporated in the military hierarchical structure but not considered as real soldiers. 'Extended family civilians' will be permitted, if they are likable, into the sphere of shared or borrowed trust that I mentioned in the introduction as an indispensable prerequisite for interviews based on confidence. In my case, being a political sociologist or a social anthropologist with a former career in development affairs was an additional factor to the required confidence building.

'National development' was, like 'national security', an all-important ingredient in the Latin American military stability and security theses until the mid-nineties of the former century. During long decades, Latin America has been the continent of political soldiers and military politicians. As a stabilising force, as 'disinterested arbiter', as a protecting power of the constitution, as guardian of national development, the military constantly intervened in political matters. In general, military ideology is ingrained with strong identification with the fate of the nation, the prevalence of order and especially the protection of the integrity of the state, and national strength and development.

Military and guerrilla are used to command but also to report and assess in unambiguous terms. Once enjoying the confidence of your soldierly partner, he or she will 'report' on situations in the past and 'assess' results of former actions. Of course there are always some institutional taboos that are to be respected: the fear that somebody will wash the dirty linen in public, for instance. However, when the interviewee is convinced of the researcher's reliability and after a sort of 'institutional permission' is given, then it becomes pretty easy to start with in-depth interviews about substantive matters.

While interviewing generals and guerrilla *comandantes* about the more military aspects, I did not find remarkable differences between both kinds of soldiers. In both categories things like honour and leadership prevail. In both cases, the soldierly component of the life history is the important backbone of interest. Where differences occur, they refer to institutional loyalties. The military's loyalty is to the armed institutions, the military´s pride is honourable conduct and the military´s fear is the immense (but undisclosed) taboo of being qualified as violator of human rights afterwards. The guerrilla's loyalty is ultimately the political motivation and the political line of his or her politico-military organisation or umbrella organisation.[17] The guerrilla's pride is that of the 'good revolutionary' and his or her fear is to be ostracised as a traitor or a 'counterrevolutionary'.

I could confirm this never extensively analysed pride-and-fear syndrome when examining the Central American post-conflict period and starting a new research project on the revolutionary generation of Cuba, the motherland and eternal icon and beacon during some fifteen guerrilla wars in other parts of Latin America and several decisive military interventions in the Horn of Africa and Southern Africa. In Cuba, El Salvador and Nicaragua one can observe the interesting transition of former guerrilla leaders in national politicians (in Cuba and Nicaragua, the actual president, and in El Salvador the actual vice-president is a former key guerrilla leader) and high ranking military. In Nicaragua, all army commanders after 1979 and in Cuba all senior four- and five-star generals after 1959 (the year of the revolutionary triumph) have a guerrilla officers' background. Their real prestige is not expressed by their most senior army rank but by their former guerrilla officers' grade (lieutenant, captain, first captain and major). A rear admiral is a 'Captain'. Ernesto Guevara was 'Major [*Comandante*] Che Guevara'; Fidel Castro is still the '*Comandante-en-Jefe*' (Chief Commander). Presidents, vice-presidents and cabinet members are referred to as '*Comandante*'. In these cases, their loyalty and pride-and-fear structure are determined by their belonging to the party organisation, the successor organisation of their former guerrilla movements.

There are two not conclusively resolvable weak points related with this interview style. The first point is the difficult problem of the reliability of the information. Self-justification and distortions of the individual memory are of course always dangers while performing in-depth interviews on sensitive themes where honour, failure and remorse can be interwoven. I only suppose that all documented history and published documentation were previously consulted. That, at any rate, will avoid conscious historical misinformation being introduced. The only, but not completely satisfactory solution, is to extend the quantity of interviews to such an extent that they cover at least one or several interviews by different persons about the same details. The second point is the question of when to stop collecting information. I use a kind of saturation philosophy: when the same information and interpretation is being confirmed by several independent sources, you can assume that you reached a certain kind of 'data mature-ness'. Both points discussed here make this interview style somewhat vulnerable. On the other hand, when oral history is the only source of data there is no methodological alternative.

My last comment concerns the personal relations during and after interviews. I always provided a listening ear, but I never expressed a personal judgement while interviewing. If asked about my private opinion, I always answered that I was only trying to understand how difficult decisions were being made and far-reaching actions were performed during difficult situations. Sometimes it felt as if I were asked to perform a kind of ambiguous therapeutic role of the researcher-interviewer. I am sure that some of the motives of granting me several of my interviews were the effort to present a sympathetic portrait of men and women acting under difficult, if not heroic, circumstances. On the other side, maybe with the exception of the interviews of the Surinamese military leaders, I scarcely encountered someone boasting or being self-important. Especially during my Cuban interviews, I only noticed a spirit of modesty.

At the end of nearly all my 160-plus interviews with army leaders, generals, former military ministers and guerrilla leaders there was always one subject tacitly and if casually discussed: was it worth it?; the implementation of the coup and the persecution of political opponents; the long-lasting military governments; kidnappings and executions during the guerrilla operations and the atrocities committed during the counterinsurgency operations. This all refers to the future standing of the military and the guerrilla leadership and the apparent personal necessity of account to the future generations.

Notes

1 Published as Kruijt (1994). I wrote about the military governments in the post-Velasco period (1975–1980) in Kruijt (1991).
2 Published as Hoogbergen and Kruijt (2005).
3 Published as Kruijt and Van Meurs (2000) and Balconi and Kruijt (2004).
4 Published as Kruijt (2008).
5 Foreword to Kruijt (1994: viii), written by the generals Miguel Ángel de la Flor, former minister of foreign affairs (1972–1976) and Ramón Miranda, former minister of education (1975–1976) and chief of staff (army) (1981).
6 Quoted from '¿Qué ejército necesita el Perú?', *Revista del Centro de Instrucción Militar del Perú*, March–April 1962 (emphasis in the original).
7 Suriname has a population of 500,000; 350,000 live in the country's capital Paramaribo. In the Netherlands there live another 350,000 Dutch-Surinamese.
8 Published as Kruijt and Maks (2004).
9 However, both a Belgian journalist and I were consulted by an intermediary about the possibility of ghost-writing Bouterse's memoirs.
10 Maroons are the descendants of slaves who escaped from the plantations during the period of slavery in Suriname. These fugitives organised themselves in tribes in the interior, in villages around the Maroni river, the border river with French Guyana. In Suriname there are five Maroon tribes: the Ndyuka, the Saramaka, the Matawai, the Kwinti and the Paramaka.
11 Before the December murders, the Surinamese military dreamed of a socialist revolution and established diplomatic relations with Cuba. In 1983, however, the Brazilian military invited themselves to a 'dialogue' with an offer of military cooperation and a clear message: 'Either they [the Cubans] are going out or we [the Brazilians] are coming in.' Out the Cubans went.
12 In 1994 the then minister of defence in Guatemala granted me permission to interview the 12 generals (ten brigadiers and two division generals) in active service. I interviewed also around 15 retired generals, most of them prominent counter-insurgency commanders.
13 El Salvador is probably the only country in the world where two adversaries, both poets, first conducted peace negotiations with one another from opposite

sides of the table and then collaborated on the publication of a collection of poems: *comandante* Fermán Cienfuegos (Eduardo Sancho), one of the five 'historical leaders' of the Frente, and university rector David Escobar Galindo, speech writer of and advisor to President Cristiani.

14 I even became aware of the significance of former love relations. In one case the lover of a Guatemalan guerrilla *comandante* explained me the complicated mixture of war and love relations between the members of the constituent politico-military organisations of the national guerrilla organisation. In Nicaragua, complicated love relations between male and female *comandantes* and commanding officers influenced public and private affairs for decades.

15 Kruijt (2008: 45).

16 Kruijt (2008: 53).

17 For instance the *Organización del Pueblo en Armas* (ORPA), founded in 1971 and in 1982 merged with other politico-military (= guerrilla) organisations into the umbrella organisation *Unidad Revolucionaria Nacional Guatemalteca* (URNG) and after the peace in 1996 constituted as a political party in 1999. Or the *Ejército Revolucionario del Pueblo* (ERP), founded in March 1972 and integrated in 1980 with four other guerrilla movements into the umbrella organisation *Frente Farabundo Martí de Liberación Nacional* (FMLN), after the peace in 1992 constituted as a political party in the same year.

References

Balconi, J. and Kruijt, D. (2004) *Hacia la reconciliación. Guatemala, 1960–1996*, Guatemala: Piedra Santa.

Hoogbergen, W. and Kruijt, D. (2005) *De oorlog van de sergeanten. Surinaamse militairen in de politiek* [*The Sergeant's War: The Surinamese Military in National Politics*], Amsterdam: Bert Bakker.

Kruijt, D. (1991) 'Peru: Entre Sendero y los militares', in D. Kruijt and E. Torres-Rivas (eds) *América Latina: Militares y sociedad*, San José: FLASCO, Tomo II, 29–142.

Kruijt, D. (1994) *Revolution by Decree. Peru 1968–1975*, Amsterdam: Thela Publishers (Thela Latin America Series).

Kruijt, D. (2008b) *Guerrillas. War and Peace in Central America*, London: Zed Books. Also published as *Guerrilla: Guerra y paz en Centroamérica*, Guatemala: F & G Editores, 2009 and as *Guerrillas: Guerra y paz en Centroamérica*, Barcelona: Icaria Antrazyt Series # 308 [Paz y Conflictos].

Kruijt, D. and Maks, M. (2004) *Een belaste relatie. 25 Jaar ontwikkelingssamen-werking Nederland – Suriname, 1975–2000* [*A Troublesome Relation. 25 Years of Development Cooperation between the Netherlands and Suriname, 1975–2000*]. The Hague and Paramaribo: Ministry of Foreign Affairs and Ministry of Planning and Development Cooperation, 2004, part 1, part 2 (report 20361 # 13 to the Permanent Commission for Development Cooperation, Parliament of the Kingdom of the Netherlands and to the Permanent Commission for Development Cooperation, National Congress of Suriname) (the report was serialised by the Surinamese journal *De West* between 3 and 9 March 2004).

Kruijt, D. and van Meurs, R. (2000) *El guerrillero y el general. Rodrigo Asturias Julio Balconi sobre la guerra y la paz en Guatemala*, Guatemala: FLACSO.

Concluding remarks

Celso Castro and Helena Carreiras

Using qualitative research methods, the contributors to this volume have put forward an important and diversified set of reflections about their research experiences in military contexts. Drawing upon these experiences, we shall now present some general conclusive comments, as well as concrete suggestions that may assist other researchers.

While the various chapters are reasonably diverse with regard to topics covered, national/organizational contexts studied and research designs implemented, there is one aspect pertaining to the research setting that makes the experiences comparable in a rather special way: the institutional isomorphism of military organizations. Beyond national and inter-service variation, military organizations possess a number of structural and cultural features that are shared cross-nationally. Military personnel are usually subject to similar processes of professional socialization and follow, in their daily lives, a normative framework based on discipline, obedience, *esprit de corps* and hierarchy. This explains why certain traits and challenges of qualitative research developed in military contexts can be shared by researchers from different national backgrounds; and also why researchers working in this field are particularly sensitive to evidence that contradicts or departs from the archetypical pattern, creating the need to develop context-sensitive analyses.

In general, one crucial issue for the success of the research is to gain access and formal clearance from the competent authorities. This is often achieved through the mediating influence of relevant or trustworthy individuals within the military establishment, from whom the researcher *borrows trust* (to use Dirk Kruijt's expression). Yet, having access and obtaining formal authorization does not guarantee, on its own, the successful completion of the research. From then on, achieving the research goals will depend on the way in which the researcher establishes interactions in the field and how the research is perceived by the participants. This idea emerges clearly from Alejandra Navarro's work, where access is understood as a critical, dynamic and flexible process through which the researcher has to engage in a process of permanent negotiation. Furthermore, as she well documents, the nature and characteristics of this process will determine many other choices throughout the development of fieldwork.

This picture can change if the researcher is an insider, a service member or a civilian working in the military or in defence-related functions. Nevertheless, justifying the research and negotiating access remain critical. Some of the experiences described in this book account for this situation and further explore the relational and ethical implications of particular ties between the researcher and the military organization. Langer and Pietsch clearly identify the dilemma:

> for those researchers who work outside the military (e.g. at civil universities), field access is often difficult to gain; and for those who work inside the military (e.g. at research institutes of the armed forces), scientific independence to choose research topics and methods autonomously and publish the research results freely may be restricted by institutional demands.

While admitting that this dilemma is not specific to research in military contexts (although it might emerge there *in extremis*), the authors provide a convincing argument for the importance of implementing a participatory methodology in the context of contract research, where results may affect the participant's life quite directly and extensively.

This shows that if some of the previous observations apply to qualitative research in general, research in military contexts has specificities that need to be taken into account. First and foremost, in the daily routines of research, it is necessary to learn how to deal with the uniqueness of an institution strongly marked by hierarchy and discipline, which develops a unique ethos and *esprit de corps*. Civilian researchers will often be perceived as part of a category – 'civilians' – which is not self-explanatory. It is vital to understand that military identity is based, to a large extent, on a detachment from characteristics perceived as civilian, which is clear in the construction of symbolic borders between the *military world* and the *outside world*. This is a focal point in the contributions by Celso Castro and Piero Leirner who document, with abundant empirical evidence, the effects of this symbolic scheme in military institutions, as well as on the conduct and results of research. Particularly important is Castro's underlying idea that dualities such as *civilian/military* or *friend/enemy* are not descriptive terms but structural categories within the military worldview and pertaining to a logic which is both situational and relational. Therefore, these are historical categories and, as such, subject to redefinition and transformation.

Some of the chapters in this book bring additional elements to complicate the opposition between those polar categories. Among them is the idea put forward by Charles Kirke that there are degrees of 'insiderness' and that the advantages and challenges that accompany them may also vary in different contexts within the same overall research. In the same vein, Dirk Kruijt's research experiences led him to understand that 'strangers' or outsiders to the military universe might be the object of multiple sub-categorizations; that a distinction is frequently made between the general public, as explicitly distinct from military and security affairs, and what he calls 'extended family civilians' of military and security significance: diplomats, historians, social and political scientists, who will, under certain conditions, be allowed into the sphere of 'borrowed trust'. Such conditions might relate to micro interactional patterns of confidence building, but also to macro patterns of civil–military relations in a country or historical juncture. For instance, as Delphine Deschaux-Beaume demonstrates in her chapter, under the concrete historical situation of the end of the French draft, or the unique citizen–soldier connection in Germany, the researcher might well be considered a *mediator* between the military sphere and society. In Deschaux-Beaume's case these are the academic and scientific circles. Recognition from and ties with society become increasingly important when organizational changes foster the interpenetrability between the military and civilian spheres. Research experiences are thus affected by varying patterns of civil–military relations.

One shared conclusion in the various contributions to this volume is thus that there are advantages and disadvantages to being an insider or outsider to an experience or organization by virtue of one's status characteristics, and this depends enormously on context.

Researchers must also be prepared to face a significant degree of external control over their work. Without falling into a conspiratorial view, it is probable that their steps will be carefully observed. This is well illustrated by Piero Leiner´s notes on the military's attempts at having absolute control over the anthropologist, by Carreiras and Alexandre's comment on gendered differences on control and surveillance and by Haddad's and Vuga and Juvan's remarks on the risks of lack of control by researchers over the choice of respondents to interviews. Therefore, it is essential that researchers are ready to permanently reflect upon their insertion into the field, and on how they and the research are presented and perceived, knowing that such perception might vary in the course of the research.

While this type of self-conscious posture is a common trait to all the chapters in the book, Liebenberg makes it the centre of his argument, underlining the advantages of an auto-ethnographic perspective. The researchers' personal experiences and trajectories impact their overall motivation for engaging in research, choice of topics, methodological strategies and ways of gathering and treating data. Exposing and revealing the researcher's self may thus be seen as a way to raise the transparency and accountability of the research.

With regard to gender issues, it is commonly recognized that despite a general tendency towards increasing equality in terms of gender, the military still preserves a dominantly masculine ethos. It is not at all clear whether this has a positive or negative impact on research – it can indeed have both – but it is an aspect which undoubtedly requires constant attention and monitoring. For instance, referring to the researcher's gender, Deschaux-Beaume describes the positive aspects of being a young woman researching a military environment and the way this status characteristic helped create empathy and rapport with the interviewees. Others have stressed the drawbacks and constraints for females conducting research in military environments. As Carreiras and Alexandre argued, the important point here is to understand that the research experience will be framed by a variety of factors, including the gendered identities of the researcher and the participants, the research design and the gendered nature of both the research context and topic. If there is no neutral point outside the gender system from where un-biased research can be conducted, it should also be recognized that the salience of gender is contingent. Therefore, the question of how gender matters can only be answered in the frame of a situated and context-sensitive evaluation.

Building trust and respect is, as much as gaining access and obtaining formal credentials, a decisive element for the success of research. Maintaining anonymity and confidentiality of the information given to the researcher is usually vital in a world such as the military, where people's actions are more visible and controlled than in other contexts. This is a problem that remains after the conclusion of the research project, and through the publication of its results. It is important to understand that sometimes the dissemination of information might be detrimental to the participants.

Until now, we have referred to the military in general, underlining common characteristics. However, although this broad category is used by all researchers in this book, it should not be naturalized; rather, it should be understood beyond its apparent simplicity and homogeneity. There are national, hierarchic and generational differences that may be decisive. One particular aspect calls our attention: from the researchers' perspective, important differences might arise with regard to feelings they develop towards their research object. Independently of how much the search for neutrality is pursued, there remains a significant difference between establishing rapport with service members who have been involved in combat and extreme life-or-death situations, and with those who have a dominantly bureaucratic career, which differs very little from that of civil servants who occasionally participate in field exercises. Likewise, contact with members of the military who have been involved in political repression or human rights violations might raise considerable challenges from an ethical and relational point of view.

As a general suggestion, the experiences collected in this book show that it is advisable to stand firmly in an explicit researcher's role, avoiding judgements and trying to understand the participant's actions and motivations.

Finally, we hope that this book brings to the fore a set of questions and suggestions that are important for future qualitative research in military studies. If research methods are understood not only as mere instruments, but as the loci of a reflexive journey towards a deeper understanding of the social world, acknowledging other researchers' experiences is certainly a useful mean towards achieving this goal.

Index

Printed by Publishers' Graphics Kentucky